T0297582

DEBUGGING EMBEDDED AND REAL-TIME SYSTEMS

The Art, Science, Technology, and Tools of Real-Time System Debugging

ARNOLD S. BERGER, PhD

University of Washington Bothell, Bothell, WA, United States

Newnes is an imprint of Elsevier
The Boulevard, Langford Lane, Kidlington, Oxford OX5 1GB, United Kingdom
50 Hampshire Street, 5th Floor, Cambridge, MA 02139, United States

Notices
Knowledge and best practice in this field are constantly changing. As new research and experience broaden our understanding, changes in research methods, professional practices, or medical treatment may become necessary.

Practitioners and researchers must always rely on their own experience and knowledge in evaluating and using any information, methods, compounds, or experiments described herein. In using such information or methods they should be mindful of their own safety and the safety of others, including parties for whom they have a professional responsibility.

To the fullest extent of the law, neither the Publisher nor the authors, contributors, or editors, assume any liability for any injury and/or damage to persons or property as a matter of products liability, negligence or otherwise, or from any use or operation of any methods, products, instructions, or ideas contained in the material herein.

Library of Congress Cataloging-in-Publication Data
A catalog record for this book is available from the Library of Congress

British Library Cataloguing-in-Publication Data
A catalogue record for this book is available from the British Library

ISBN: 978-0-12-817811-9

For information on all Newnes publications
visit our website at https://www.elsevier.com/books-and-journals

Publisher: Mara Conner
Acquisitions Editor: Tim Pitts
Editorial Project Manager: Rachel Pomery
Production Project Manager: Prem Kumar Kaliamoorthi
Cover Designer: Mark Rogers

Typeset by SPi Global, India

Contents

Preface

Why write a book about debugging real-time systems? Good question, I'm glad you asked. A lot has been written about debugging real-time systems or embedded systems, but what has been written has not, to the best of my knowledge, been collected into one resource, such as a book.

After having taught embedded system design for many years, I've come to the conclusion that we are failing as teachers because our students can write a program in assembly, C, C++, C#, some Arduino dialect, or Verilog, and get their program to compile. However, if problems crop up, as they invariably do, students lack the diagnostic skills to analyze the problems and, in a systematic way, zero in on the possible causes, and then find and fix the bugs. I hope to address this issue in the chapters to come.

What I observe with depressing regularity is that students take the "shotgun approach." Try a bunch of changes at once and hope for the best. Even more disturbing, rather than try to find and fix a problem, students will just throw away their code or their prototype and start all over again, hoping beyond hope that will fix the problem.

You might assume when the students of today become the engineers of tomorrow and are totally immersed in product design, they will have developed the debugging skills they need to do their job in the most effective manner. I've learned that assumption does not hold true.

Before I became an academic, I worked in industry, creating design and debug tools for embedded systems designers. In particular, I designed and led teams that designed logic analyzers, in-circuit emulators, and performance analyzers. These were and in many cases still are complex instruments designed to solve complex problems. Just learning to effectively use one of these instruments can be a chore that many engineers don't feel the desire to invest the time required to learn.

Maybe you've been there yourself. Do you do a mental cost/benefit analysis to invest the time to wade through a set of manuals,[a] or just dive in and hope for the best? One of the most

[a] No kidding, the user manuals for the HP 64700 family took up about 1 m on a bookshelf.

brilliant engineers I ever worked with, John Hansen, made this observation that came to be known as Hansen's Law, which says:

If a customer doesn't know how to use a feature, the feature doesn't exist.

So, as vendors of these complex and expensive debugging tools, we certainly own a good part of the problem. We have not been able to effectively transfer technology to a user in a way that allows the user to take full advantage of the power of the tool to solve a problem.

Here's another example. I remember this one vividly because it led me to think a whole new way about technology and how to transfer it. We'll come back to it in a later chapter, but this is a good time to introduce the problem. It involves logic analyzers. For many years, the logic analyzer has been one of the premier tools for real-time system analysis. There's some evidence that dominance may be changing, but for now, we'll assume the logic analyzer still holds a position of prominence.

Suppose you are trying to debug a complex, real-time system with many high-priority tasks running in parallel. Stopping the processor to "single step" through your code is not an option, although many excellent debuggers are task-aware, so they may be able to single step in a particular task without stopping other tasks from running at full speed.

The logic analyzer sits between the processor or processors and the rest of the system, and records in real time the state of every address bit, data bit, and status bit output from the processor, then inputs to the processor as they occur in real-time. Once the buffer or recording memory is full, the engineer can then trace through it and see what exactly transpired during the time interval of interest.

But how do you define the time interval of interest? Your memory buffer is not infinitely large, and your processor is clipping along at 10 s to 100 s of millions of bus cycles every second. This is where the logic analyzer really shines as a tool. The user can define a sequence of events through the code, very much like a sequence of states in a finite state machine. In some logic analyzers, these states can be defined in terms of the high-level C++ code with which the engineer is accustomed to programming.

If the user gets the sequence of states correctly defined, the logic analyzer will trigger (capture) the trace at just the right time in the code sequence to show where the fault occurs. Here is where it gets interesting. At the time,[b] the HP (now Keysight) logic

[b]This was in the mid-1990s, so clearly the situation is rather ancient, but the point still holds today.

analyzers had relatively small trace buffers but very sophisticated state machines. The design philosophy was that the user didn't need a deep trace buffer because she could zero in on exactly the point where the problem occurs. In fact, the state machines on the logic analyzers were eight levels deep. Here's an example of how you might use it. Refer to Fig. 1 below.

This example is three levels deep and each level had many options in terms of defining the state, or the number of times a loop might run. What we discovered was that our customers rarely if ever tried to set up the trigger condition beyond two levels. What they didn't like about the product was that our trace buffer was too shallow (5000 states). They preferred simple triggering with deep memory to complex triggering with shallow memory. This was pretty consistent with each customer visit we conducted.

What's the point? Instead of using the powerful triggering capability of the logic analyzer, the engineers we spoke with preferred to take the path of least resistance. They preferred to manually wade through a long trace listing to find the event of interest rather than learn how to set up a complex trigger condition. In other words, trade off using a complex tool for a less capable but easier-to-use tool.

You could argue that the engineers were just lazy, but I think that's the wrong perspective. I'm sure that if we, the tool designers, could have invented a more intuitive and user-friendly interface

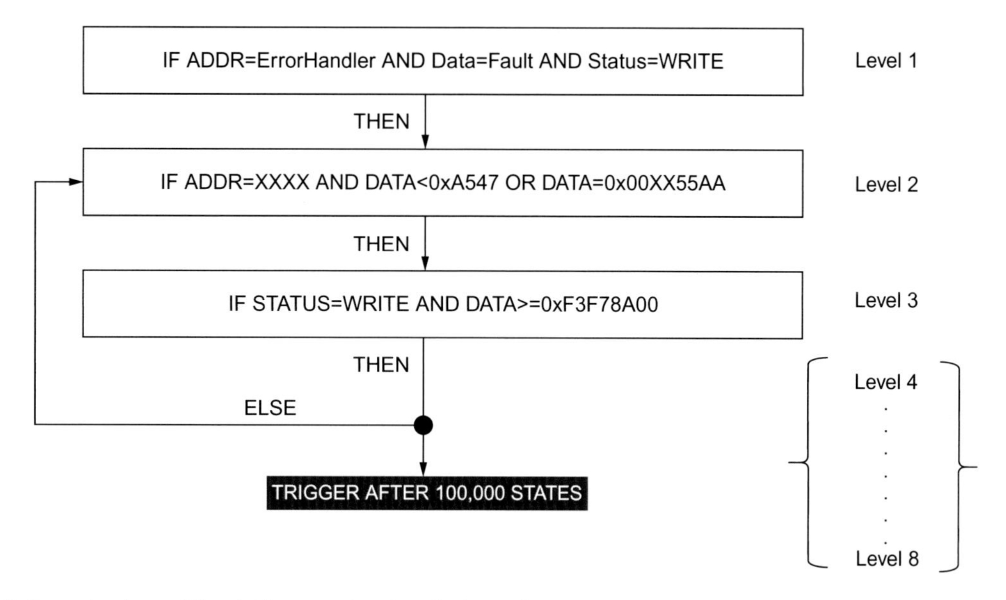

Fig. 1 Example of a multilevel trigger system in a logic analyzer.

between the engineer trying to solve a difficult debugging problem and the tool itself, the engineer would have gone for the best solution every time and his debugging skills would have improved in the process.

Why did I bring up this example? I wanted to mention it up front because debugging real-time systems is often very difficult and engineers need to use complex tools in order to bring high-quality products to market in a timely manner. If this book contributes to learning how to use tools or sensitizes the reader to dig deeper into the user manual, then this book will have served its purpose.

Let's discuss this book. The initial focus is aimed at the student. Not just the ones who want to enter the field of embedded systems design, but rather all Electrical Engineering, Computer Science, or Computer Engineering students who wish to improve their skills in the debugging of their designs. Also, you would be able to casually mention during a job interview that you've taken some effort to go beyond just doing projects by also being able to bring defect-free projects to completion. You've just expressed to the interviewer that you are entering the job market with a skillset beyond what your graduating peers might have.

For the experienced engineer who is already a practitioner and wishes to hone his or her skillset, I hope this book will provide you with a roadmap to tools and techniques that you may not be aware of, or to more efficient ways to solve the problems that seem to crop up on an ongoing basis.

In researching and writing the book, I decided that application notes and white papers were the very best sources of information on specific categories of bugs. Having written more than a few of these articles myself, I am pretty confident that this was a good decision.

If you think about it, it becomes obvious. Companies are constantly polling their customer base for design and debug problems that invariably appear as the technology advances. These customer problems are the driving force for the creation of tools that will solve the problems, or at least point to the source of the problems.

Once the tool is invented, its potential value must be explained to the customer base, which leads to presentations at conferences, technical articles in industry publications, and application notes that link the problem, the source of the problem, the tool, and the solution in a way that the engineer can internalize and justify purchasing to upper management.

While these articles are clearly self-serving for the companies generating them, they are also valuable resources that provide the best up-to-date and practical information that engineers need. For me, they became my principal resource for this book.

We'll start by examining the debugging problem itself. What is the nature of real-time systems that makes debugging so unique? At this point you might say, "Duh, its real time." But that's only a part of the problem. For a large fraction of embedded systems, the fixed hardware, reprogrammable hardware, firmware, operating system, and application software will probably be unique, or at least a large fraction of the design is unique.

Even with products that are evolutionary, rather than revolutionary, there may be many new elements that must be integrated into the overall design. So, the problem is more than simply real time versus not real time. The problem is really the number of variables in the system AND the fact that the systems must run in real time as well.

After we scope the problem, we'll turn our attention to an overall strategy of how to debug hardware and software. We'll look at best practices and general strategies. Also, it will be useful to consider testability issues as well because debugging a system that wasn't designed to be debugged in the first place can be a challenge.

From strategies, we'll turn our attention to tools and techniques. We'll look at some classic problems and look for methods to solve them.

Next, we'll look at what kind of support silicon manufacturers provide in the form of on-chip debugging and performance resources to help their customers bring new designs to market.

The final section of the book will cover serial protocols and how to debug them. According to some experts in the field, more and more of the debugging problems are related to serial data movement and less are lending themselves to classical debugging techniques.

I've tried to make this an easy read rather than a pedantic tome. I've put a lot of personal anecdotes in because they can help make a point and they're fun to read and fun to write. I took my lead here from Bob Pease's classic book, *Troubleshooting Analog Circuits* [1]. Many senior engineers remember Bob's column in Electronic Design Magazine, "What's all this...."

His book and these columns were classic reads and I am unashamedly borrowing his conversational style for this book as well. As an aside, my favorite "What's all this...." column was about his analysis of these ultrapricy speaker cables that were cropping up, claiming audio advantages over simple lamp cord wires. Bob does a rigorous analysis that is both fun to read and educational at the same time. More to the point, he pretty well puts the audio superiority claims to bed.

I've also noticed that I haven't strictly adhered to keeping subject material isolated in the appropriate chapters. You'll find that some examples may appear in different chapters. That's by design, rather than me having a "senior moment." It comes from teaching. I often repeat and review material in order to place it in a different context, rather than simply lecturing on it and moving on. So, if you see a discussion of common real-time operating system (RTOS) bugs in the chapter on common software defects and then run into it again in the chapter on debugging real-time operating systems, don't say I didn't warn you.

Arnold S. Berger

Reference

[1] R.A. Pease, Troubleshooting Analog Circuits, Butterworth-Heinemann, 1991 ISBN 0-7506-9499-8.

1

What's the problem?

Software engineers write code and debug their code. Their platform may be a PC, Apple, Android, Linux, or another standard platform. Electrical engineers design circuits, run simulations, create printed circuit boards, stuff the boards, and then run confidence tests. They will also typically need to debug their designs.

If the software issues and the hardware issues are kept separate from each other, then the bugs that may be found in each domain can be complex and challenging, but are generally bound by the domain in which they reside.

Now, let's really muddy the water. Let's bring these two domains together and allow them to interact with each other. Suddenly, we have multiple opportunities for interaction and failures to occur:

- There is a defect in the software that is independent of the hardware.
- There is a defect in the hardware that is independent of the software.
- There is a defect in the system that is only visible when hardware and software are interacting with each other.
- There is a defect in the system when it must execute an algorithm with a finite window of time.
- There is a defect in the system when multiple software applications are running concurrently and vying for the limited hardware resources under a control program (real-time operating system, or RTOS).
- There is a defect in the system that only occurs in rare instances, and we cannot recreate the failure mode.

I'm sure that you, the reader, can add additional instances to this list.

Let's consider the traditional embedded systems development life cycle model shown below in Fig. 1.1. This model has been presented so many times at every marketing demonstration that it's fair to question whether it represents reality. Let's assume for a

Debugging Embedded and Real-Time Systems. https://doi.org/10.1016/B978-0-12-817811-9.00001-6

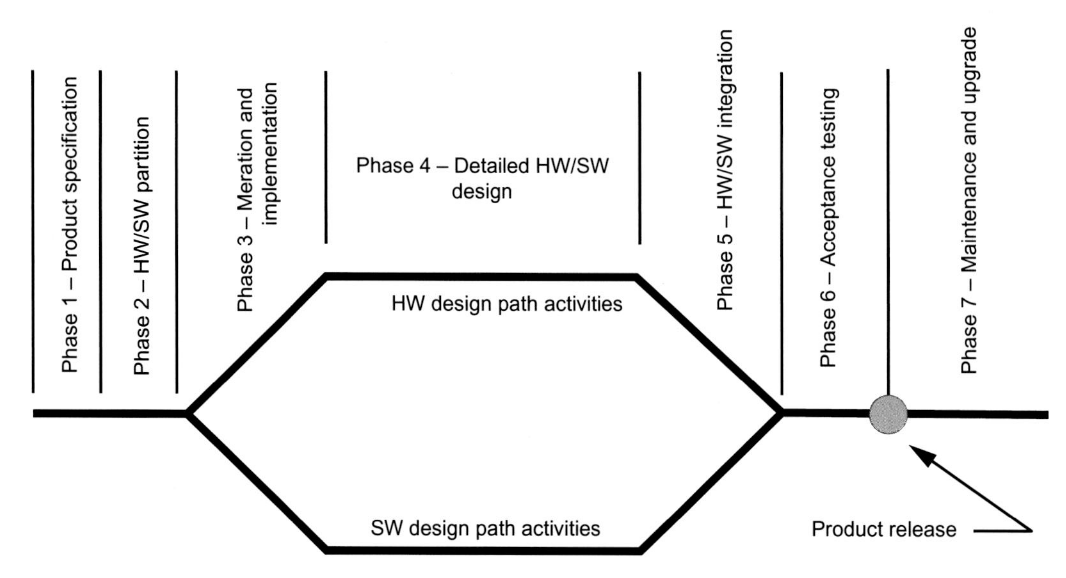

Fig. 1.1 Embedded system development model [1].

moment that it does represent how a large fraction of new embedded designs are approached.

We might assume that everything is fine until we reach phase 3. Here, the software team and the hardware team go their separate ways to design the hardware and the software to the specifications developed during phase 1 and phase 2.

We may certainly question whether defects could have been introduced in these initial two phases that will not become apparent until the specifications are physically implemented.

We'll ignore that aspect for the moment. Broadly speaking, this bug can be a marketing requirement that, for various reasons, is impossible to achieve. This marketing "bug" is a specification defect. Many will argue that this is not a bug in the true sense of the word, but rather a shortcoming in the product definition phase. However, finding the defect and fixing it can be every bit as challenging and time consuming as finding a bug in hardware timing or software performance.

In the traditional sense, the defects are created in phase 3 and phase 4. The hardware designers work from the specification of what the physical system will be and the software designers work from the sense of how the system must behave and be controlled. Both these dimensions are open to interpretation and there may be ambiguities in how the specification is interpreted by each team or by each individual engineer.

You might argue this won't happen because the teams are in contact with each other and ambiguities can be discussed and resolved. That's valid, but suppose that the teams don't realize an ambiguity exists in the first place.

Here's a classic example of such a defect. In fact, this is one of the top 10 bugs in computer science. It's called the Endianness Problem (https://en.wikipedia.org/wiki/Endianness).

This ambiguity occurs because smaller data types such as bytes (8-bits) or words (16-bits) can be stored in memory systems designed to hold larger variable types (32-bits or 64-bits). Because all memory addressing is byte-oriented, a 32-bit long data type can occupy four successive byte addresses.

Let's assume we have a 32-bit long word at the hexadecimal address **00FFFFF0**. Using byte addressing, this long word occupies four memory addresses, 00FFFFF0, c, 00FFFFF2, and 00FFFFF3.[a] The next long word would begin at byte address 00FFFFF4 and so forth within this long word.

We see in Fig. 1.2 that there are two ways byte addresses may be arranged within the four-byte, 32-bit long word. If we are following the Big Endian convention, then the byte at hexadecimal address 00FFFFF1 would be Byte 2. But if we are following Little Endian convention, then Byte 2 would occupy memory address 00FFFFF2.

Suppose that this memory location was actually a set of memory-mapped hardware registers and writing a byte to a

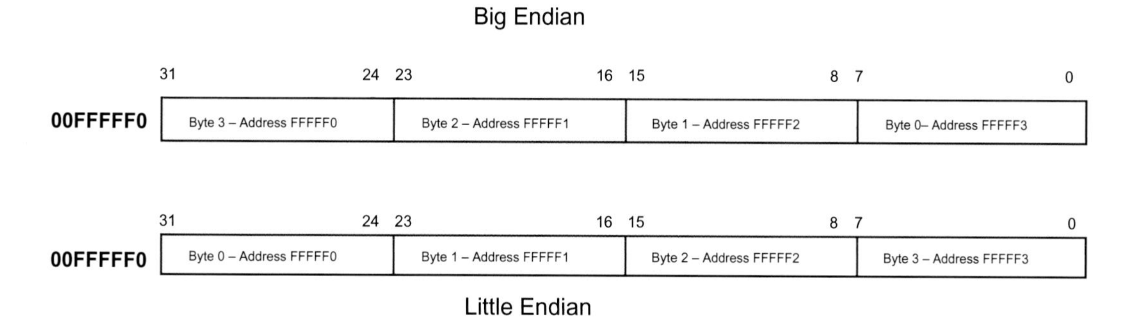

Fig. 1.2 Illustration of the ambiguity in a byte address caused by different assumptions about the Endianness of the processor to memory interface.

[a] There is another set of possible defects that might happen if our long word started at an address not easily divisible by four, such as 00FFFFF1, but we'll delay this discussion for a later time.

specific address is supposed to cause the hardware to do something. Clearly, if the hardware designer and the software designer have different assumptions about the Endianness of the system and have not explicitly discussed them, then a defect will be created by omission in the formal specification to declare the Endianness of the system.

If you're lucky, you'll catch the error in simulation or in a design review. However, the odds are just as likely that you won't catch the error and the bug will become visible when commands are sent to the wrong portion of the 32-bit hardware register.

Even worse, engineering teams may be geographically separated from each other. I once comanaged a project that was a joint effort between my Hewlett-Packard division, located in Colorado Springs, Colorado, and a semiconductor company's R&D team located in Herzliya, Israel. Sure, we had regular status meetings via conference calls, but there were many times when the key engineer was not available or we weren't able to resolve all the issues during the call. Issues fell through the cracks despite our best intentions to keep communicating with each other, and there were several product defects resulting from communications breakdowns.

Unless you are going to design hardware and software through some hybrid model of Extreme Programming (https://en.wikipedia.org/wiki/Extreme_programming) with the hardware designer and software designer sitting together, one writing C code and one writing Verilog HDL code, you can expect that defects will arise that will eventually need to uncovered, analyzed, and repaired.

If the defect is in the hardware, the hardware is an application-specific integrated circuit (ASIC), and the design has gone to fabrication in a silicon foundry, then the cost of fixing the hardware may be prohibitive and the defect will need to be fixed by a software workaround (if possible).

This workaround can cause a chain reaction because in the extreme (Murphy's Law), the overall system performance is impacted because software ends up doing work that hardware cannot. We've fixed the hardware defect, but we've introduced a real-time performance defect in the process.

Most of the defects will become apparent during the integration phase of the project when the hardware and software are first brought together. The concept of a hardware/software integration phase is one that is often considered to be either false or misleading. Best practices dictate that the integration of hardware and software should be continuous during the development phase with frequent testing of modules against each other.

One way to implement such a system is to use HW/SW coverification tools, such as Seamless from Mentor Graphics in Beaverton, Oregon. An excellent series of articles [2] provides a good basis for understanding the technology of coverification. Briefly, in coverification, software is continuously tested against the virtual hardware, which exists in a logic simulation. The coverification tools provide the interface between the instruction set simulator (ISS) for the software by creating a bus functional model (BFM) of the processor. The BFM is the glue logic between the ISS and the logic simulator. It accepts the machine level instruction and converts it to the address, data, and status bit activity the processor would execute if it were physically present.

The output side of the BFM plugs into the hardware simulator. This can be a Verilog simulator if the hardware is in an FPGA or an IC model, or a system logic simulator if the rest of the system is made up of discrete logical blocks.

Thus, using a coverification design approach, whether you actually use the coverifications tools, you should expect the integration phase (if it exists at all) to go more smoothly and with fewer bugs to be found and fixed.

When we introduce the third dimension of time constraints and a fourth dimension of real-time multitasking, the opportunity for the introduction of additional defects rises exponentially. Let's consider one such set of issues, hard real time versus soft real time [3].

In a hard real-time system, a missed deadline is fatal: the system doesn't work, the plane falls out of the sky, people are killed. Here we have a knife-edge decision to make. Either we fix it or the embedded system is unusable.

What is the best strategy to fix the flaw? Is the processor underpowered? Is the memory system too slow? Should we run a faster clock? Maybe. These are all issues with which the hardware team would grapple. Is software causing the problem? Maybe. Are hardware and software both causing the problems? Very possible. What about the software tools (see below)?

A hard real-time failure is serious because it is unlikely there is a workaround, an easy fix that can get around the flaw. Workarounds fix errors that can't be fixed any other way, but they never make things better.

Let's consider the case of a well-known laser printer manufacturer. Their design philosophy was to wring as much performance out of the processor as possible. They had a really good engineer who examined the code that was output from the C compiler and handcrafted it to make it as fast and efficient as possible. This sometimes means rewriting sections of C code in assembly language to eliminate some of the overhead inherent in high-level languages. They realized that if they had a hard real-time failure, the problem had to originate with the hardware because the software was already optimized.

The lesson here is that once you've determined the reason your system has hit the performance wall and the cause is hardware, the most likely path to improve performance is to respin the hardware. This may involve:

- Choosing a faster processor.
- Running a faster clock with an existing processor.
- Improving the memory to processor interface.
- Adding additional custom ASIC hardware.
- Adding more processors to offload the problem processor.
- Some combination of the above.

My point is to emphasize the importance of accurate analysis before taking any further action. This would likely involve:

- Assessing the problem.
- Analyzing the problem.
- Deciding on a course of action.
- Get buy-in (very important).
- Resolve it.

Here's another example (Although the problem is a soft real-time issue, which is a topic covered in Chapter 2, it is relevant to our current discussion). A manufacturer of RAID arrays was convinced that their RAID control card required a faster processor. They spoke with their local sales engineer from their current microprocessor vendor and were assured that the new processor coming down the pipe had twice the throughput of their current processor and was code compatible. However, the new processor had a different footprint (pad layout) than their current one, necessitating a redesign of their current version of the RAID controller.

The company went ahead and did a respin of the control board and was devastated to find the throughput improvement was a factor of about 1.15, rather than the 2 × improvement expected. After further analysis, they found the problem was a combination of several poor software architectural decisions; switching to the new processor was not the solution. The company almost went out of business because of their failed analysis.

Let's move on to look at soft real-time failures. In contrast to a hard real-time failure, a soft real-time failure is one where the system takes longer to complete a task than is desirable or necessary for the product to be competitive in the marketplace. A good example is the laser printer. Let's assume the marketing specification calls for the printer driver to output 20 pages per minute (20 ppm).

The HW/SW integration phase goes smoothly, but the printer's output is 18 ppm. Clearly, it doesn't meet specs. The design team (and marketing) must decide whether to fix the bug, which might take weeks or months to correct, or release the product with lower performance and risk being noncompetitive in the marketplace.

This was discussed earlier when we looked at the situation of a defect in a custom IC that caused the correction to be made in software. Software, though very flexible, is not as fast as dedicated hardware, and perhaps this is the root cause of the poor performance. Alternatively, it may be a poorly matched compiler with the chosen microprocessor. When I worked with embedded tools vendors who supported Advance Micro Devices (AMD)-embedded microprocessors, there was a 2:1 performance difference between the lowest performance C compiler and the highest performance C compiler. A 2:1 difference is the same as running the processor at half the normal clock speed.

I was involved in another interesting soft real-time defect situation. My company, Applied Microsystems Corporation (now defunct), created a hardware/software tool called CodeTest. The tool was later sold to Metrowerks when AMC went out of business. CodeTest worked by postprocessing the user's code and inserting "tags" in various places, such as the entry and exit points of functions. These tags were designed to be low-overhead writes to a specific memory location that could be collected by the hardware portion of the tool.

Unlike a logic analyzer, CodeTest could continuously gather tag data, time stamp it, compress it, and send it to the host computer where the data was being collected; a runtime image of the software execution could be displayed. It was, and still is, a very neat tool for software performance analysis.

Back to the bug. We were demonstrating CodeTest to a telecommunications manufacturer. They had a soft real-time problem they were trying to find. Their product performance was no longer adequate for the market and they were trying to decide if they could upgrade the existing product or if they needed to design a new one.

We set up CodeTest and started to gather statistics. One function seemed to be taking up a lot more of the CPU cycles than any other function and no one on the software team would admit to being the one who developed the code.

After some searching, the company's engineers discovered that the function was written by an intern who was doing some board testing and wrote a high-priority test function that blinked a light on the back of one of the PC boards in the device. This code was never intended to be part of the released product code but somehow was added to the code build for the released product.

Once the code image was repaired and the errant function was removed, the telecomm manufacturer's product was back in business. We lost a sale, but they bought us a nice dinner instead. How do we categorize this bug? It was certainly a soft real-time defect, but the root cause was the company's source code control and build system. However, it is unlikely that without CodeTest, the engineers would have gained the insight they needed to see the product defect and fix it.

Once we add multitasking into the mix via an RTOS (real-time operating system), the possibility of system defects becomes even more likely. Before we discuss RTOS and defects, a few words about the way an RTOS operates are in order. Any computer operating system generally has the same purpose. It wants every independent application running concurrently to be well isolated from every other application and, for any individual application, it should appear to have all the systems resources available to it all the time. It makes writing programs infinitely easier for any processor doing more than a few simple concurrent tasks.

I'm writing this book on a laptop computer running Windows 10 as its operating system. If you count the number of programs that I have open, plus all the software running in the background, there are more than 30 applications all running at the same time. Well, sort of running at the same time. Windows is mostly a round-robin operating system. Every active program gets a time slice and there is an internal timer that allocates time slices for each application. These time slices are generally fast enough so that all the programs appear to be running simultaneously, though only one[b] application is actually running at a time. Round-robin operating systems are perfectly acceptable and desirable for desktop or laptop computers, but are totally inadequate for real-time systems because real-time systems have time applications that require prioritizing CPU cycles based on the criticality of their time constraints. A task with a hypercritical time window can't wait until its next turn in the queue. It must preempt all other less-critical tasks and run as soon as it needs to run.

A real-time operating system is a priority-driven O/S. The more important tasks have priority over the less-critical tasks. If a lower

[b]The assertion that one program at a time is running gets fuzzy when we use CPUs with multiple cores, but in order to make the most effective use of these cores, the O/S must be able to manage them. This is not an easy task. Some applications have been written specifically to take advantage of multiple cores, but most applications do not have this ability.

priority task is running while the higher priority task wakes up and needs to run, the lower priority task is temporarily put in suspension while the higher priority task takes over and runs to completion, or is preempted by an even higher priority task.

What might go wrong? One of the simplest defects is CPU starvation. The CPU is so heavily loaded that a low priority task is basically starved of CPU availability and never gets to run to completion. If you want to see a manifestation of this phenomenon, although not quite the same thing, try running Windows 10 on a computer with less than 4 GB of memory installed.

You'll see delays when you move the mouse or strike a key on the keyboard as the processor struggles to give all the programs the time they require. Because there is so little memory installed, the hard drive, which is 10,000 × slower than RAM, must take over and hold the programs as they are being swapped in and out of memory.

Another classic defect involving an RTOS is called priority inversion. Suppose a low priority task has requested a system resource under control of the RTOS and has been granted the use of the resource. This could be a timer, a communications channel, a block of memory, etc. Suddenly, a high-priority task wakes up and needs the use of the resource. While it is possible for a higher priority task to preempt a lower priority task, when system resources are involved, this becomes more problematic. The lower priority task must relinquish the resource as soon as possible, but it can't immediately relinquish the resource because that could leave the resource in a metastable state.

Thus, the RTOS waits until the resource is free before turning it over to the waiting task. In the interim, a medium priority task that does not require the resource is able to preempt the lower priority task and run. This is priority inversion. The highest priority task has been preempted by a lower priority task. In other words, the two priorities have been inverted.

The most famous of these system failures occurred on the Mars Pathfinder Mission in 1997. The Sojourner vehicle stopped communication with the Jet Propulsion Lab in Pasadena and began to exhibit systematic resets that wiped out all the data collected that day.

The story was recounted in a famous Dr. Dobbs interview with Glenn Reeves, the Flight Software Cognizant Engineer for the Attitude and Information Management Subsystem, Mars Pathfinder Mission [4]. The title of this article says it all:

REALLY REMOTE Debugging: A Conversation with Glenn Reeves

Quoting Glenn Reeves:

> *In less than 18 hours, we were able to repeat the problem, isolate it to an interaction of the pipe() and diagnose it as a priority inversion problem, and identify the most likely fix.*

The point of this remarkable story is that even after your hardware and operational software are running correctly, the new dimension added by the introduction of a resource control and scheduling program (an RTOS) can lead to yet another opportunity for defects to be introduced into your system.

One way to avoid this cross-coupling problem is to partition the problem among multiple independent processors. Ganssle [5] discusses this as an approach to improve productivity and defeat the productivity reduction caused by large programs and the need to maintain communications among many programmers.

It may seem counterintuitive at first, as engineers are trying to optimize our designs, but throwing transistors[c] at a problem may be the easiest way to design a multitasking embedded system without resorting to a real-time operating system. Not every design lends itself to this approach, but many do. Look at today's automobiles. Each one is a mobile computer network with nearly 100 unique microprocessors and microcontrollers. Many of these processors have just one task to perform.

My favorite automobile example is the luxury car with a unique capability. If a driver closes the door from the inside when all of the other doors and windows are closed, a single itty-bitty microcontroller in each door will crack open the window just as the door is about to shut. As soon as the door is shut, the window goes back up. It happens in seconds.

The purpose of this processor is to avoid the air pressure burst on the driver's ear when the door is shut. Cool, yes, but why do this? Because we can. I could design a circuit using a programmable logic array or discrete components and replicate this circuit, but the solution would not be as simple as a 4-bit microcontroller costing less than $0.10 USD and probably requiring fewer than 100 lines of assembly code.

As we wrap up this chapter, it is instructive to reflect on what I've tried to say here and its relationship to designing embedded systems. Due to the often-chaotic state of software development, a great deal has been written about methods for designing software (less so about designing hardware). A systematic method

[c]This is hardware designer's jargon for adding hardware.

of specifying, coding, and testing software should be an engineering discipline and not an art form.

So, why write a book about finding and fixing bugs when the whole idea is to design hardware and software that is bug-free in the first place?

Here's what I've concluded:

- We're still not a place where, for whatever reason, software and hardware can be designed without the introduction of defects.
- Electrical engineering, computer science, and computer engineering students, by and large, have never been taught a systematic process for finding and fixing defects.
- Tools and techniques for finding and fixing bugs often go unused because engineers don't know they exist or don't make the effort to learn the tool.
- Some combination of the above.

In the next chapter, we'll tackle the first problem. You know you have a problem, so how do you find it and fix it?

References

[1] A.S. Berger, Embedded System Design, CMP Books, Lawrence, KS, ISBN:1-57820-073-3, 2002, p. 2.
[2] J. Andrews, https://www.embedded.com/design/debug-and-optimization/4216254/4/HW-SW-co-verification-basics–Part-1—Determining-what—how-to-verify, 2011.
[3] X. Fan, Real-Time Embedded Systems: Design Principles and Engineering Practices, Newnes, ISBN: 978-0-12-801507-0, 2015, p. 6.
[4] J. Woehr, REALLY REMOTE Debugging: A Conversation with Glenn Reeves, https://www.drdobbs.com/architecture-and-design/really-remote-debugging-a-conversation-w/228700403, April 2010.
[5] J. Ganssle, The Art of Designing Embedded Systems, Newnes, Boston, 2000, p. 37 0-7506-9869-1.

2

A systematic approach to debugging

In this chapter, I am speaking directly to the students and faculty. I would hope that experienced engineers are well versed in the techniques that I'll be discussing, and new engineers are more at the student's end of the spectrum. To the student reading this, please excuse the fact that I seem to be directing the content toward your instructor because I am referring to you in the third person.

I thought I'd introduce this chapter with something that Joe Decuir,[a] an IEEE Fellow and an affiliate faculty member of my division at the University of Washington Bothell, sent me.

The six stages of debugging

1. That can't happen.
2. That doesn't happen on my machine.
3. That shouldn't happen.
4. Why does it happen?
5. Oh, I see.
6. How did that ever work?

Now, back to the problem at hand. After observing my students struggle time after time trying to find and fix problems with their hardware, their software, or both, I decided to share with them some "best practices" that I've learned over the years to approach the problems of:

1. Observing a bug.
2. Being able to reproduce the bug.
3. Hypothesizing the cause of the bug.
4. Testing the hypothesis.
5. Achieving high confidence that the hypothesis is correct.
6. Making the correction.
7. Retesting to validate the fix.

[a] https://en.wikipedia.org/wiki/Joseph_C._Decuir.

Debugging Embedded and Real-Time Systems. https://doi.org/10.1016/B978-0-12-817811-9.00002-8

In short, it didn't work. If I gave a quiz or question on a test, I'm confident that most students would get the right answer. But suppose that you're struggling with a design problem, or a circuit that isn't doing what you designed it to do, would you know where to begin? Would you make matters worse by "shotgunning" the circuit problem?[b]

> *Shotgun debugging is the debugging of a program, hardware, or system problem using the approach of trying several possible solutions at the same time in the hope that one of them will work. This approach may work in some circumstances while sometimes incurring the risk of introducing new and even more serious problems.*

In my experience, it never works. Shotgun debugging of software may just be a time sink, but for hardware it is the kiss of death. Why is this? For starters, printed circuit boards are not designed for continual heating and reheating cycles on the pads or traces. Delicate parts are also not designed to withstand repeated heating and reheating cycles.

So, what do students do when confronted with a hardware bug? They start to change parts and hope that will fix it. Most of the time they just make matters worse. They destroy the parts that they only have one of. Ditto with the PC board. Time goes by and they are no closer to solving the problem and deeper in the hole. More telling, they have no idea where their original starting point was, so they can't get back to it even if they wanted to. In the extreme, they will start over again, hoping that by just redoing the board or rewriting the code, the bug will resolve itself. This doesn't work either.

So, if they actually know better, why don't they follow the best practices that I've laid out for them? I suspect they'd rather try something quickly and hope for the best than spend the time required to set up a systematic debug process to find the defect.

At this point, I'm going to take a short detour and caution the reader about one debug technique that is fraught with danger. When we teach our incoming EE students their first circuits class, we provide them a lab kit of parts that includes a solderless breadboard. Fig. 2.1 is a photo of a typical solderless breadboard.

If you were an incoming student taking our Circuits I and II classes, you would build simple DC and low frequency AC circuits with these breadboards and within that context, they sort of work okay. Provided that you are working with signals greater than 10 mV or so and frequencies under 1 kHz, the solderless breadboard will enable you to do rapid prototyping.

Depending on how neatly you do the wiring, you (more or less) can do your lab experiments, get reasonable results, and pass the

[b]This definition encapsulates the method quite well [1].

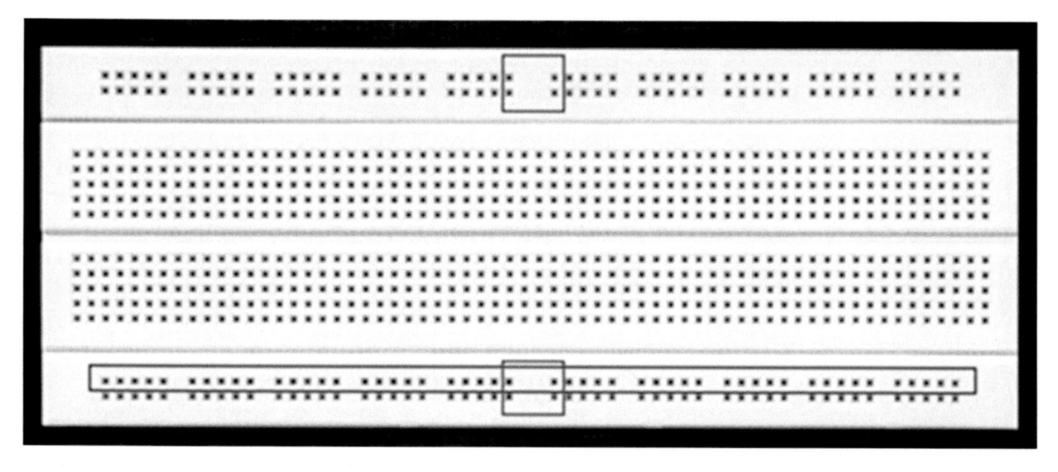

Fig. 2.1 A solderless breadboard (https://www.auselectronicsdirect.com.au/arduino-solderless-breadboard-840-points?gclid=EAIaIQobChMI2O6Nt5S-2gIVmnZgCh2f2gLZEAQYASABEgKlqPD_BwE). Vertical rows of five dots (two are highlighted) are connected together. Horizontal rows (one row is highlighted) are connected together. Inserting 22-gauge solid core wire makes the necessary connection. The two *blue* squares indicate points where the horizontal rows are segmented into eight separate buses. Jumper wires are required to connect them to form four continuous strips for power and ground.

course. The same holds true for simple digital circuits. As long as the circuits are simple and low frequency, then the logic chips work as promised, although close examination of fast rising and falling edges shows significant overshoot, undershoot, and ringing, even at nearly DC switching speeds.

The situation becomes dangerous when you use the solderless breadboard as your go-to debugging tool for all circuit design problems and for rapid prototyping of higher-sensitivity or higher-frequency circuits. At that point, the solderless breadboard becomes a negative productivity tool. It causes more time to be wasted as the student attempts to figure out why their circuit isn't working. You may have wired it correctly, but the signal either looks terrible or it is buried in ground loop noise. Here's how I learned not to use solderless breadboards.

I was a relatively new engineer at the Colorado Springs Division of Hewlett-Packard. I was doing my first real circuit design and I was prototyping part of the circuit on a solderless breadboard. Resistors, capacitors, wires, and transistors were all standing up in a three-dimensional circuit.

A senior engineer peeked over the wall to my cubicle, saw what I was doing, walked in, and pushed down on the circuit with his palm, squashing everything together. Never saying a word, he walked out. That was my lesson.

When is it OK for you to use a solderless breadboard? I won't freely admit this, except under duress or in this tell-all book, but I will on occasion resort to using a solderless breadboard. Almost all the time, it is when I don't understand something in the data sheet regarding the part I'm trying to design into the circuit, and I want to confirm some functional feature of the part. Reading the data sheet is often futile, so If I don't have the simulation model for the part, I'll try out a sample part on a solderless breadboard in order to confirm that I understand how it works. If I really get stuck, I'll call an applications engineer at the company that manufactures the part.

Having just warned you of the dangers of using a solderless breadboard, I couldn't resist adding the cover photo, Fig. 2.2, from Bob Pease's classic book, *Troubleshooting Analog Circuits* [2]. As you can see, he does not heed my warnings about the use of solderless breadboards. I suspect that this photo was staged for effect. Here's what Bob, one of the gurus of circuit design, had to say about solderless breadboards:

> *I didn't even think about solderless breadboards when I wrote my series*[c] *because I see them so rarely at work. They just have too many disadvantages to be good for any serious work. So, if you insist on using these slabs of trouble, you can't say I didn't warn you.*

Who's at fault here?

I believe that educators are remiss in not teaching how to find and correct errors in design with the same rigor as we teach the basic principles of engineering circuit design or writing software. A survey conducted in 2005 at the Embedded Systems Conference in San Francisco[d] indicated that:

> *...debugging is the most time-consuming and costly phase of the software development lifecycle, with a majority of respondents citing debugging as the most significant problem they encounter.*

Therefore, if debugging is the costliest and most time-consuming phase of the development life cycle, why aren't engineers taught the best ways to debug code (or hardware, or Verilog)? Here are some thoughts:

[c]Bob Pease wrote a regular series, *Pease Porridge*, in Electronic Design magazine. He died in a car crash in June 2011.

[d]https://www.businesswire.com/news/home/20050413005208/en/Survey-Reveals-Debugging-Time-Consuming-Phase-Software-Development.

Fig. 2.2 Cover photo from the book, *Troubleshooting Analog Circuits*, by Robert A. Pease. Photo used by permission of the publisher, Elsevier, Ltd.

- Debugging is viewed as a negative; a bug is a flaw. It's better to design code without bugs in the first place.
- After you teach the basics of coding, or designing circuits, there isn't enough time left in the quarter or semester to teach debugging as a discipline worthy of study.
- The best debugging tools have yet to be invented because current tools aren't up to the challenge.
- Debugging has not been designated as an academic discipline worthy of study, scholarship, and research, so it just isn't taught.
- Best practices are handed down from senior engineers to junior engineers and are not disseminated outside a tightly knit design team.

A bug of my own

Let's start now to try to define an overall strategy to approach the identification of the root cause of a defect in hardware (or software) and then how to fix it. In order to create the strategy, we'll start with a case study. I'm going to describe a simple embedded system that has a bug, one that I designed for my course at the University of Washington Bothell, B EE 425, *Introduction to Microprocessor System Design.*

This particular design was for the laboratory part of the course. I wanted students in the class to learn how to use a logic analyzer to look at the interplay between hardware and software. The board contained a 4 MHz Z80 CPU, 32 K ROM, 32 K RAM, a single 7-segment display, and some glue logic, including a Dallas Semiconductor (now Maxim) micromonitor IC that handled power-on reset, watchdog timer, and a reset push button. This is the 8-pin IC located just behind the Z80-CPU.

The Z80 was chosen because it is an excellent teaching example of how a microprocessor works. The student can see the signals on the processor's I/O pins in real time while the code is executing. Everything that the processor does is unambiguously visible to the logic analyzer. There is no cache and no input pipeline queue; just the basic clock, address bus, data bus, and status bus. Yes, the Z80 is old and slow, but if you want to give a lecture on a typical timing diagram for a generic microprocessor, a logic analyzer trace of a Z80 bus cycle unambiguously matches the lecture diagram.

The seven-segment display was added in a later revision of the board because students invariably complained that their board was dead. The display and an appropriate test ROM chip would show them that the board was working.

After building and loading a test run of three boards, I wrote test code that would exercise the board and allow me to look at all the bus signals with both a logic analyzer and an oscilloscope probe. The scope showed the fidelity of the signals and the 34-channel logic analyzer allowed me to see almost all the Z80's address, data, and status pins (Fig. 2.3).

All the test boards turned on. I next wrote the code that ran the display. Against all rules of good design practice, I wrote a software timing loop so that the display would run slowly enough for a human being to observe it. I didn't have the room to add a separate timer to the board, so the software loop was my only way to slow the system down without adding more hardware. The three boards ran my code well and the displays continuously scrolled b-E-E-4-2-5.

Buoyed with confidence, I ordered 20 more boards and enough parts to stuff them. Because I didn't want to build 20 boards myself, I organized a Saturday morning soldering class for the EE students who'd never soldered a board before. Before turning them loose on the good boards, I had them watch several YouTube videos on soldering. Next, I gave them some practice PC boards and assorted parts to solder. When the students were able to demonstrate that they "sort of" knew what to do, we turned them loose. Several student assistants with soldering experience circulated and also kept an eye on the soldering work.

Fig. 2.3 Lab experiment board for a microprocessor course. The 40-pin connector connects the board to a LogicPort Logic Analyzer.

When a board was completed, we plugged in the display ROM to see if it worked. It was gratifying that about 15 of the 20 boards turned on right away. The other five did not, so we put those aside. Later, I carefully examined the five that did not turn on. Two had easy-to-fix soldering issues and worked after I repaired them. Three looked like the soldering was good—no bridges, no cold solder joints—but the boards still would not turn on.

Rather than discuss how I found the bug, I'm going to describe the systematic process that *I teach students to follow* to find and fix the defect. Write down what you know and what you've observed; you need a paper trail. This is like placing rocks along a poorly marked hiking trail so you can find your way back. Also, creating a paper trail will help you organize your thoughts and create some hypotheses that you can then test.

a. Observed: Three Z80 boards out of a batch of 20 failed to turn on using the EPROM test code that activates the display.

b. Known:

- **b.1.** The code and EPROMs work in other boards.
- **b.2.** Visual inspection shows good solder joints and no solder bridges on any boards.

c. Possible causes:

- **c.1.** Power to ground short circuit.
- **c.2.** Manufacturing defects in boards.

c.3. Bad clock.

c.4. Shorted or open address, data, or status signals.

c.5. Overheating pads or traces during soldering.

c.6. Parts improperly inserted in sockets.

c.7. Bent IC pins not in socket.

c.8. Problem in seven-segment display circuit.

c.9. Other possible causes I haven't thought of yet.

At this point, let's take a breath and review where we are. We've identified a minimum of nine possible hypotheses to start investigating. Of course, there could be more. We haven't even considered a software bug, but the simplest line of testing appears to be a hardware defect because these are new boards built by inexperienced engineering students loading and soldering their first boards.

The plan of attack will be to check the boards in more detail, starting from the simplest possible cause and working up to full-scale board debugging. If necessary, connect a logic analyzer and observe all the relevant signals.

In my notebook (or your lab notebook), I created a list of each test I was going to perform, what I expected to see, and what I observed.

Tests to perform:

1. Using an ohmmeter, check for a short between the power and ground.
   Expected: Should measure some resistance greater than 0 Ω.
   Measured: 150 Ω.
2. Check power to board.
   Expected: All Vcc pins have +5 VDC.
   Measured: Every Vcc pin measured 5.25 VDC.
   Comment: Used 5 V wall supply from lab kits, 1A DC current limit.
3. Check the parts. Are they properly inserted? Are they overheating? Do a thorough visual inspection, and then remove all parts from the socket, inspect the pins, and reinsert.
   Expected: All parts have been properly oriented and inserted in their respective sockets.
   Observed: All parts were properly inserted and no parts were excessively hot.
4. Check clock signal.
   Expected: 4 MHz, 0–5 Vpp (Vpp = peak to peak voltage) square wave observed at crystal oscillator output and at Z80 clock input.

Observed: Clock looked good at the crystal and at Z80c clock input (pin 6).
Comment: Used lab oscilloscope grounded to ground test point on the board.

5. Using a logic analyzer, check for open or shorted address, data, or status traces.
Expected: Trace listing should agree with source code binary output.
Comments:
    a. Used state mode.
    b. Triggered logic analyzer trace on exiting reset (~RESET goes high).
    c. Concerned about software timing loop filling buffer.

Observed: Code executed properly and entered software delay loop, filling buffer. Problem does not appear to be shorted or open traces. Still did not observe display operation.

6. Check to see if code exits software delay loop.
Expected: Code will properly exit the loop, indicating problem is somewhere else.
Comments: Trying to eliminate software delay loop by reconfiguring logic analyzer trigger circuit to sequentially trigger. Set condition "A" to be the first address in the software delay loop and then set condition "B" to be the first address after the software delay loop.
Observed: LA did not trigger. Software never exits the delay loop.

Here is the first appearance of a clue. It doesn't seem to be a hardware bug, but we can't rule it out just yet. More tests are necessary. After scratching my head for a while, and rereading the help menu for the logic analyzer, I modified the trigger condition so that state "B" was any address outside the range of the software delay loop that the code is failing to properly exit.

Before going any further, I need to write down what I just observed so that I have a data point to which I can return. Next, I repeated the test with the logic analyzer, but now with the modified trigger conditions. Set trigger sequence so condition "A" is the first address in the delay loop. Condition "B" is any address outside the delay loop.
Observed: Logic analyzer triggered.
Comment: Trace listing showed that the RESET went low and the processor began to reexecute the initialization code.

Now I knew where the problem was; I just didn't know why it was happening. Once in the delay loop, the processor would reset itself and start over. It never exited the loop to drive a new value to the display.

The only way to generate a hard RESET signal is to push the RESET button, which I definitely was not doing, or for the RESET signal to be generated by the hardware. The suspect now was the micromonitor IC from Maxim. In addition to the power-on reset, reset button, and power supply monitoring, the micromonitor also has a watchdog timer. Could that be triggering and asserting RESET?

I thought I understood how the part worked, but I went back to my data sheet and carefully read the section about the operation of the watchdog timer. According to my design, I left an input pin labeled TD, for timer delay, unconnected. When this pin is left unconnected, the watchdog timer will time out in 600ms if it does not see an input on its STROBE pin. When the timer times out, it generates a RESET signal. I was using one of the status signals from the Z80 to prevent the watchdog timer from asserting RESET, but that signal was not active when the software was in the timing delay loop.

In an incredible validation of Murphy's Law, the watchdog timer timeout value and my timing delay loop were just about equal. There was just enough variability in the exact timeout delay in the DS1232 micromonitor chip so that most of the chips worked, but some did not. The simplest way to test this hypothesis was to take a working board and a defective board and swap the micromonitor chips. If the failure followed the chip, I knew the cause of the problem and rereading the data sheet gave me the solution.

7. Swap the micromonitor chips in a working and nonworking board.
   Observed: The failure follows the DC1232 chip.
   Proposed solution: Rewrite the delay loop code or change the watchdog timer timeout.
8. Connect a jumper wire from the TD input of the micromonitor chip to the Vcc. This should increase the timeout from 600 to 1200ms.
   Observed: A board with this failure mode now worked properly.

   Once again, I stopped and documented what my tests had shown. I knew what was causing the problem and wrote down what I did to verify that I had indeed found the root cause of the problem. I also wrote down a possible software-only solution that would negate the need to solder a jumper wire on the 20 boards.
9. Rewrite the software delay loop to continuously output a signal to the STROBE input on the micromonitor chip so that I would not time out and generate any more errant RESET signals.

Expected: All boards should now work without the need for a hardware jumper wire.
Observed: All defective boards now worked.
What are the lessons to take away from this exercise?

1. Stop, think, and plan before attacking the problem:

The exercise of first writing down what I knew, what I suspected, and what I thought I should check was critical. It provided a framework for isolating the defect before I started to change anything.

Also, by writing my plan down, I forced myself to slow down and not jump in. If the problem was much more complex and there were multiple engineers involved, this first step would likely involve a meeting and a brainstorming session to capture our ideas and then structure them into a coherent plan.

There can be a lot of steps in a plan and, once we start to debug the system, we may have to revise our roadmap. That's okay and to be expected. As we get deeper into identifying a problem, new clues may present themselves and we would need to look into other areas that were not part of the original plan.

Another part of the preliminary work is to identify as many variables as you think you are going to need to track. This can easily become a laundry list of compiler versions, makefiles, linker command files, data sheets, etc. This is probably overkill, but when you are under the gun and trying to fix bugs so you can meet your product release deadline, having all the information at hand is priceless.

Of course, you can keep your notes in electronic form. I'm old school. I'm used to having a lab notebook to capture ideas and make notes.

If you've ever done any repair work on your car, you've probably invested in a factory service manual. These manuals are designed by the automotive engineers for the garage or dealer mechanics who have to fix a car. Go to any section in the manual and you will find a flow chart of the recommended process for isolating a mechanical or electronic failure.

A flow chart is an alternate form of the list of steps, but has the added advantage of allowing you to view your debugging plan as a roadmap and also list what you want to do at decision points. Perhaps it is overkill, and I can honestly say I've never had a problem that required this type of approach, but it could be beneficial. Let's just leave that in the "might work" category.

2. Thoroughly read and understand the data sheets:

In the book "*Debugging: The 9 Indispensable Rules for Finding Even the Most Elusive Software and Hardware Problems*" [3], one of the key points author David Agans makes is to understand your system. This means thoroughly and carefully reading all the data sheets as well as all the code you are using that you didn't write yourself, or code you wrote so long ago that you no longer remember how it works.

Agans further makes the point that you should carefully read all the comments in any software that is linked to the defect, or might be linked, or even has no apparent linkage. He gives one example where he was debugging an embedded processor system and the code was written in assembly language. One of the on-chip registers was getting corrupted. Here's the author's description of the problem:

We were debugging an embedded firmware program that was coded in assembly language. This meant we were dealing directly with the microprocessor registers. We found that the B register was getting clobbered, and we narrowed down the problem to a call into a subroutine. As we looked at the source code for the subroutine, we found the following comment at the top: "/ Caution—this subroutine clobbers the B register. */"*

Agans devotes an entire chapter to this point. You should read everything, particularly the comments in code that you did not write, or you wrote a long time ago (anything older than 2 weeks is a long time). Only then can you eliminate the critical variable that you don't fully understand the system. Here's a summary of his nine indispensable rules for debugging hardware and software bugs. I'll be returning to these points over and over in this chapter, and subsequent chapters to follow:

- Understand the system.
- Make it fail.
- Quit thinking and look.
- Divide and conquer.
- Change one thing at a time.
- Keep an audit trail.
- Check the plug.
- Get a fresh view.
- If you didn't fix it, it ain't fixed.

While I'm on the subject of the need to completely understand your design, here's another relevant real-world example. Just prior to my departure from Hewlett-Packard's Logic Systems Division (LSD) in the mid-1990s, I was managing a project that was in the initial phases of product definition. We had a product idea, but we weren't given the go ahead to start developing the product. The code name was

"Farside." Not because of the Gary Larson comic strip, but because we were looking beyond conventional tools.

As we defined it, Farside was going to be the penultimate debugging tool for embedded systems. We were aiming this debugging tool at engineers who needed to understand already working systems that needed to be upgraded. We settled on this type of debugging tool when we were doing customer research at a large telecommunications company on the East Coast. There we met with a team of engineers who were tasked with taking an existing product that they had not developed and making it better.

All they had to go by was the standard documentation, service manuals, all the software, and all the hardware schematics. What they lacked was any documentation describing the intent of the original designers or their original design philosophy. What impressed me was their ability to read through these reams of documents and extract the design philosophy and the places where improvements could be made. What they were asking from us was a reverse engineering tool or suite of tools that would enable them to see the design at a higher level than just reading through the source code or schematic diagrams.

These superb engineers defined the way to learn about your system.[e]

3. Trust Occam's Razor:

As I'm sure most of you know, Occam's Razor is *the problem-solving principle that, when presented with competing hypothetical answers to a problem, one should select the one that makes the fewest assumptions.*[f]

Another way to state this principle is that *other things being equal, simpler explanations are generally better than more complex ones.*[f]

How does Occam's Razor apply to debugging? Let's reconsider my process chart for debugging the microprocessor board. The very first test I did on the nonfunctional boards was to see if power and ground were shorted together. That would certainly explain a dead board. Also, it was plausible that a student may have damaged a board by overheating it and causing an inner layer short. This wasn't part of my hypothesis;

[e] Logic Systems Division was shut down by HP just after we were given the green light to develop Farside. At that point, we forgot about the project and focused 110% of our energy on finding gainful employment somewhere else.

[f] https://en.wikipedia.org/wiki/Occam%27s_razor.

I just wanted to start with simple but significant tests and work my way up to the more subtle tests.

Note that I did not start with step #5: connect the logic analyzer. I proceeded from simple, global tests and worked my way toward the more complex possible problems as I eliminated the simple ones.

4. Do differential testing:

This means just change one variable at a time. Differential testing is the exact opposite of shotgunning. By changing only one variable, you can see the effect that the change had on the system. Ask the student why they don't follow this process and the consistent answer you will hear is, "*I don't have the time.*"

If you're a student reading this, you will probably agree that time is your most precious asset. You never have enough time. Therefore, you would probably sympathize with their comment. Perhaps you've been there more times than you want to admit. But if you don't have time to follow a disciplined process, why are you so willing to tear something apart, rewire it, or rewrite it, without stopping to analyze what could be the real source of the error?

It's painful to watch when a neatly laid-out solderless breadboard that obviously took quite a long time to construct is pulled apart and rewired because the student hadn't taken the time available to properly locate the flaw.

In this particular case, the student had wired the board correctly and had simply neglected to check the positive and negative power rails. If he had performed this simple test, he would have discovered that half the circuit wasn't connected to power or ground. The breadboard he was using actually split the power and ground rails in the middle of the rail. The upper four rail segments were connected to power and ground, but the lower four rail segments were unconnected. All that was required to fix the problem was four little jumper wires between the upper and lower segments of the power and ground rails. Referring back to Fig. 2.1, the two blue squares show where the power and ground buses are split. The four segments on the left side of the breadboard are connected together and the four segments on the right side of the breadboard are connected together, but there is no connection between the two halves.

In the previous example in this chapter (pg. 21), I showed this principle in step #6. Up to this point, I had narrowed the defect down to the possibility that the micromonitor IC was the leading suspect, but I wasn't certain it was the culprit because I had 17 boards that worked just fine and three that didn't. I had a pretty good idea that was the problem because I knew that the time delay and watchdog timer reset interval were about the same amount of time.

Also, after reading the data sheet for the part, I knew there was some variability in the exact time interval of the watchdog timer, so it followed that some parts might work and other parts might not work. Thus, the differential measurement was to replace a suspect part with a part that did not generate the RESET pulses.

Testing one variable at a time, and only that variable, makes good sense. Every time you simultaneously test more than one variable, you change the number of possible outcome combinations by at least 2^N, where N is the number of variables you changed.

Now you might think, "*Why is he saying 'at least'* 2^N? Isn't it just 2^N?" If there were no additional interactions between the N variables, then the number of possible outcome combinations might be just 2^N and, to clarify, this statement is saying that if you do see a change in the system with a particular combination of input variable changes, then there are 2^N-1 other combinations that might also cause the change.

However, if two or more of those variables are interdependent, or the problem stops (or a new problem begins) because of the interplay between two variables, then you are exposing your debugging session to even more possible outcomes and exponentially more time to sort them out.

Even when you do use differential testing, it is still critically important to write down your expected result and what you actually observe. Even if you debug properly, it is easy to lose track of the results of the tens of tests that you are performing.

5. Consider the best solution before moving on:

It's tempting to do a quick fix; and there are lots of "good reasons" for them including:

a. We're behind schedule.
b. The "big show" is next week.
c. The software team needs units to continue development.
d. Got to get units out into the field ASAP.
e. Engineers have to move onto other projects.

If you are a student reading this, you will have your own set of good reasons to want to do a quick fix and move on. These would likely include:

a. The project is due at the end of the week.
b. I have to study for my (name the course) final exam.
c. I'm leaving town after finals are over.

d. I'll lose my scholarship if I don't get a good grade in this class.

In order to close the book on this section, we need to return to the example of the errant Z80 boards. We've identified the problem and made a software fix in the delay loop that eliminated the watchdog timer timing out and causing a ~RESET to be asserted. Should we just move on and call it fixed?

The quick fix was to rewrite the code and reprogram the ROMs. However, there was another solution. I could change the timer's timeout interval to 1.2 s by bringing the TD input to the +5 VDC (Vcc) power plane. This involves adding a jumper wire from the TD to the +5 VDC input pin. Not a particularly hard modification to do to three boards. Might take 30 min in total. But wait. This is a lurking problem. Shouldn't I modify all 20 boards to prevent the problem from recurring? That would easily have taken several more hours.

Alternatively, I could do a redesign of the board and permanently tie TD to Vcc. Now I have to buy a new batch of boards and reload them. Time consuming and costly. Better yet, wouldn't it be nice if you could select the time interval for the watchdog timer with a permanent jumper selector? If you look back at Fig. 2.2, you can see jumper selectors in the lower left of the board. Why not add another one for the timer?

At my division of Hewlett-Packard, one of the project managers had a plastic baseball bat with the phrase "WIBNI Killer" on it. WIBNI stands for "Wouldn't It Be Nice If…" When an engineer suddenly got excited about adding more features or capabilities, the manager would bring out the WIBNI Killer and symbolically beat the engineer into submission.

Why this aside? Because if we have a window of opportunity to fix the hardware, it is reasonable to ask if we should use the opportunity to also make some design modifications or improvements. If there will need to be a respin of the board to fix our defect, perhaps we should use the opportunity to bring in something that marketing has been requesting.

The key point of this discussion is that we should not just fix the bug with the most expedient possible fix and move on. Once you know what the problem is, take time to list your possible solutions before taking the easiest path. To review, the possible solutions are:

a. Change the coding of the software delay loop.
b. Add a jumper wire from the TD to Vcc.
c. Respin the board and add a trace from the TD to Vcc.
d. Respin the board and add a selectable jumper block for three possible timer timeout delays.

In this example, with 20 boards for a student lab experiment, the simplest solution was the best one. Change the ROM code and eliminate the watchdog timer timeout issue. That's what I did in this situation, but it wasn't the only option we could have considered.

In summary, most of the time we find the problem, fix it, and move on. I'm suggesting that before moving on, you take a moment to consider more possible options. You might be surprised.

6. Log the defect:

Keeping track of the defects you've found and repaired is also a critical part of the process if you want to improve your design processes. In my simple example, I would list the root cause as two factors:

a. An incomplete understanding of how the DS1232 worked.
b. Poor software design by using a software timing loop.

What is the lesson here? That's not so easy. I knew that using a software timing loop was not good design practice, but it was a reasonable solution for a simple situation where I needed a way to show the students that a board was working properly, and their problems were elsewhere.

Would a better understanding of the micromonitor actually have prevented the bug? Maybe. Maybe I could have taken the time to calculate the exact elapsed time of the time delay loop that I wanted for the display, instead of writing the delay loop, watching the display, and figuring "that's about right," then looking at the possible variability in the timer timeout to see if there was a possible conflict.

That approach would probably have found the bug. Interestingly, I doubt that this bug would have been caught in a design review. I would have gotten some flak for the software delay loop, but I could have justified it on its merits.

The hardware flaw (if you can call it a flaw) might have been picked up because the TD input on the DS1232 was left unconnected. Open inputs are always something to check as they can be sources of system instability due to noise on the floating inputs. However, I would have justified it as an intended part of my design. I think it would have taken someone really astute to pick up this defect before I uncovered it.

Even so, having a written analysis of the bug, what you observed, and how you fixed it can be an incredible time saver at some time in the future. If you simply fix it and move on, you may forget about it and then, fast forward 2 years, there it is again. You've got this vague recollection that you've seen this before, but you can't recall what exactly it was and what you did to find and fix it. So you go through the debug process once again. You find it and fix it, but imagine how you could wow the crowd if you opened up your defect logbook (paper or electronic) and found the defect notes you made long before.

Let's close this chapter on the generalized process of debugging by summarizing the main points of the discussion.

1. Stop, think, and plan before attacking the problem:

   Write down what you've observed and what the possible causes might be. Revise this plan as necessary. Keep a written record of what you expected to see, assuming the system is working properly, and what you've actually observed.

2. Thoroughly read and understand the data sheets:

   Failure to understand how aspects of your design work is a primary reason that defects can be introduced into an embedded system.

3. Trust Occam's Razor:

   Explore the simple possibilities first, then move on to the more subtle problems.

4. Do differential testing:

   Never change more than one variable at a time, and never change several variables at once in the hope that The Force will be with you and one of those fixes will be "the one."

5. Consider the best solution before moving on:

   It's tempting to implement the quickest fix to correct the bug, but give it some thought and choose the best overall solution, not the most expedient.

6. Log the defect:

   You can't implement continuous improvement unless you can learn from your mistakes. Keep a record. A history of errors and their causes is an invaluable tool.

A final thought.

Another avenue of debugging is to rerun simulations of the circuit and see if they provide a clue to the behavior you are observing. We'll discuss using simulations in a later chapter when we talk about how to best use the arsenal of debug tools that is available to you.

I'll end this chapter with another student story. Our EE, CS, and CE students at UWB are primarily focused on earning a BS degree and then joining the workforce. A small fraction will go on to grad school to get an MS degree, and only a very small fraction will choose to go on to earn a Ph.D. Their Capstone Final Exam is the technical interview that they will likely go through with a prospective employer.

Because I was a hiring manager for many years in industry and interviewed many, many new engineers, my students tend to listen when I talk about the hiring process and only "multitask" by viewing their Twitter accounts when I talk about the subject matter of the course. When the students hand in their final project reports, I ask that they do a self-critique of their project and what they could have done better.

Easily one-third to one-half of the students describe how they just ran out of time due to debugging the hardware or the software. A typical response is "I decided to use this Arduino code I found online and ..." or "I found this App Note on the manufacturer's web site...." When I give them a project grade, I write a fairly long and detailed critique of their project and their report.

Often, I'll suggest that they just redo their project the right way and then write the report about how they followed the development process ***that we covered in class*** and used all the ***best practices*** that we discussed. No matter how badly their project turned out the first time through, they could still write an excellent and compelling project report that would influence any hiring manager to seriously consider this student.

Picture this, right there in the report is the debug write-up, just like we discussed in class. This is the report that should make it into your portfolio, not the one that was submitted for a grade. I'm reminded of my days as a high school student taking first-year physics lab. We were assigned an experiment to measure acceleration due to gravity. Of course, our experimental data taking left much to be desired and our data was wildly off, but no worry. We knew where we wanted to end up, so we simply started with the result (32 ft/s/s) and then we worked backward to create the data to fit.

References

[1] https://searchsoftwarequality.techtarget.com/definition/shotgun-debugging.

[2] R.A. Pease, Troubleshooting Analog Circuits, ISBN # 0-7506-9499-8Butterworth-Heinemann, Boston, MA, 1991.

[3] D.J. Agans, Debugging: The 9 Indispensable Rules for Finding Even the Most Elusive Software and Hardware Problems, Amacon, a Division of the American Management Association, New York, 2006, p. 11. ISBN 0-8144-7457-8.

3

Best practices for debugging embedded software

Introduction

In this chapter, we'll explore some of the best practices for debugging embedded software. I'm going to focus on embedded C and assembly language because they're what I know best, and they are typically the languages that are most tightly linked to embedded systems.

I've included a reading list of articles that I came across in my research. These articles cover many of the topics in greater detail than I can hope to cover. Also, they tend to focus on how to write bug-free code because it is always better to write bug-free code than it is to find and fix the bugs after the fact. It is also cheaper[a] to find bugs earlier than it is to find them later.

Fig. 3.1 is a graph showing the relationship between the timeline in a project development cycle and the cost to fix a bug. Notice that the curve is exponential, not linear, which makes sense: the further you are in the product design life cycle, the more costly it is to fix mistakes.[b] However, modern technology might mitigate this somewhat.

Given the prevalence of nonvolatile memory in modern embedded systems, often within the microcontroller itself, it is quite straightforward to fix defects in the field. This assumes that the device can be connected in some way to a communications link and that provisions are built into it for seamlessly upgrading the firmware. If this capability is not available, then this graph is accurate.

[a] Use whatever metric you like here: money, engineering time, missed delivery dates, resources, etc.

[b] I am writing this chapter in the midst of the Boeing 737 MAX software failure that has grounded the entire fleet of 737 MAX jets while the problem is being fixed.

Debugging Embedded and Real-Time Systems. https://doi.org/10.1016/B978-0-12-817811-9.00003-X

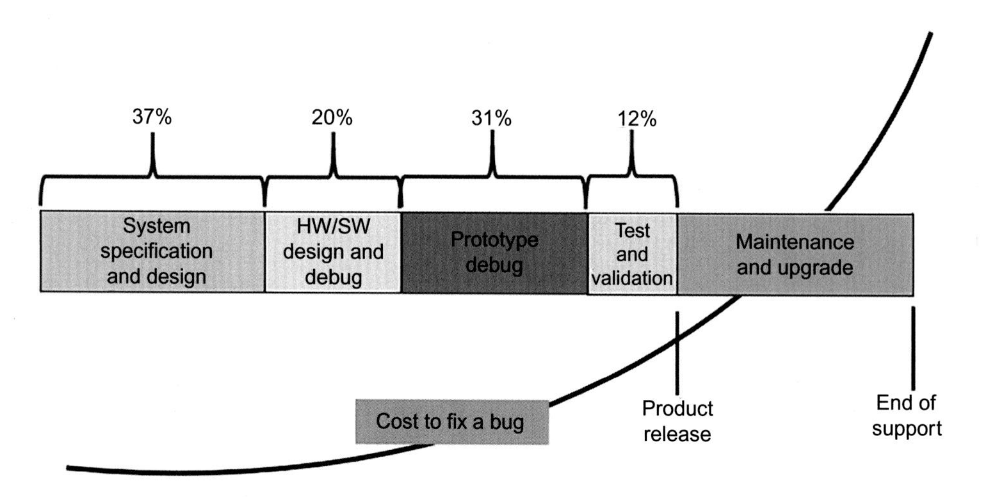

Fig. 3.1 The cost to fix a bug versus the phase of the embedded design cycle [1].

Also, note that the data show that slightly more than half the design cycle time is spent in the debug phase of the project. This data is fairly consistent across a number of articles I reviewed, although several outliers had the number as high as 80%.

While this number spans all software development, I would venture to guess that this would be a low number for embedded software because of the added dimension of HW/SW interactions.

I don't want to tread over ground that has already been covered by other authors, so let's look at the problem from an embedded perspective and proceed from there.

What makes embedded systems different?

This is a good place to start our investigation of best practices in debugging embedded systems. While embedded software (firmware, operating system, and application) share many attributes with every other flavor of software, whether it is PC-based, mobile, or mainframe, there are differences that place embedded software in a unique category. Here are some of the differences [2]:

Embedded systems are dedicated to specific tasks while PCs are generic computing platforms

Today, we could argue whether a smart phone is another form of personal computer or an embedded system. A smart phone certainly has dimensions of both the traditional embedded system, such as task-specific requirements (phone, GPS, Bluetooth communications, etc.), and with the applications that run on the phones and make them an integral part of our daily lives.[c]

Of course, there are still many, many more single-purpose embedded applications than smart phones, but we should acknowledge that the division between traditional embedded devices and modern ones is not a sharp one.

The implication of software failure is much more severe in embedded systems than in desktop systems

I mentioned the Boeing 737 MAX issues in a footnote b. Due to a system failure of some kind, more than 300 people died in 2 separate crashes. We could debate whether it was a hardware failure (the angle of attack sensor was producing faulty data), a software failure (correct data from the sensor was misinterpreted), a design failure (a critical system was not triply redundant), a specification problem (pilots were not given sufficient information about the operation of the MCAS system), a marketing failure (warning lights about the system were sold as options), a corporate failure (the plane was rushed into production to better compete with its Airbus rival), or finally, a government failure (the FAA allowed Boeing to self-certify the aircraft) [3].

Whatever the cause, the MCAS system failed and lives were lost as a result.

Other recent failure incidents involve the next big trend on the horizon—self-driving cars. In March 2018, an Uber self-driving Volvo hit and killed a pedestrian in Tempe, Arizona [4]. According to the article, the human driver failed to take control from the car to prevent the fatality.

The classic fatal software failure involved the Therac-25, a radiation therapy machine produced by Atomic Energy of Canada Limited [5]. This incident became a case study of poor software design and the lack of due diligence when it came to safety-critical software design.

At least six people were overdosed with lethal radiation when faults in the software allowed the operator to inadvertently improperly set the radiation parameters, and the safety backups, present in previous models of the Therac, were removed in model 25. We know that embedded software must not fail even

[c]It's always interesting to look at my students' faces when I tell them to turn off their phones and put them away for the duration of the class.

if lives are not at risk. It is our expectation as customers that our devices containing embedded microprocessors will just keep working. This is the main reason that most embedded microcontrollers contain watchdog timers.

I owned an inexpensive ink-jet printer a while back that would periodically reset itself. If I were a betting man, I would venture to guess that buried deep inside the control processor was a watchdog timer that tried desperately to compensate for a software bug that was forcing the firmware to go south every once in a while.

Without the watchdog timer, it would have been an unacceptable product and I would have returned it. Instead, it was a cute "feature."[d]

Embedded systems have real-time constraints

By their very nature, embedded microprocessors are control elements. They are devices that must interact with time. You would be hard-pressed to find any microcontroller, no matter how simple and inexpensive, that does not contain one or more timers of some kind.

There are generally two types of real-time dependencies: *time-sensitive* and *time-critical.*

A time-sensitive dependency is one where performance degrades gracefully with time. A laser printer produces 18 pages a minute, for example, instead of 20 pages a minute. Marketing is livid but no one is injured. The printer works. It may not be as competitive in the market as its closest rival that prints 22 pages a minute, but it still works.

A time-critical dependency is far more serious. If a required action does not take place within the allotted time window, the system fails. A plane with a fly-by-wire control system falls out of the sky. An industrial process controller runs amok. You get the picture.

Now, throw a real-time operating system (RTOS) into the pot and things get even more convoluted because the software behavior is no longer deterministic. You can't simply count CPU instruction cycles and figure out how long a particular function will take to execute.

Your software might be perfect, but the system may still fail due to a once-in-a-million sequence of I/O events. You will never find this bug by single stepping through your code or doing a code inspection.

[d]There are times when a defect is called a feature. I guess it depends upon your point of view.

In one of the best talks I ever heard at an Embedded Systems Conference, Dr. David Stewart [6] analyzed the 25 most common errors in embedded software development. Dr. Stewart starts at number 25, the least serious, ("My problem is different") and works backwards to number one, the most serious. That defect, according to Stewart, is "No measurement of execution time."

Stewart said, *"Many programmers who design real-time systems have no idea of the execution time of any part of their code."*

I think you can see why writing code without being aware of the real-time constraints of the design is a bad thing.

There are several different design projects that I give to my microprocessor systems class. One of my favorites is to design a function generator capable of output sine, triangle, and square waves at frequencies up to 100 kHz. For these EE students, this is the first real design assigned that takes them out of their comfort zone of solving problems from the end of the textbook chapter.

They grabbed an Arduino Uno or Mega and forged ahead. When they demonstrated their designs, many stopped working at 100 Hz, just 1000 times below the design requirement.

What went wrong? Even though I covered performance issues in class, the students never bothered to figure out how fast they had to output the sine values from their look-up table in order to create one full sine cycle every 10 μs.

Embedded systems are supported by a wide array of processors and processor architectures

I tried to find out how many unique microprocessor and microcontrollers there are in the world today, but I felt like I was going down a Google rabbit hole. So, let's be conservative and guesstimate there are more than 1000.

Many have adopted standard architectures such as ARM Cortex while others are totally unique. For each of these architectures, there must be a suite of software tools available to support it. At a minimum, we need an assembler and linker. Going a step further, we get into compilers, debuggers, and perhaps some additional architecture-specific tools.

Moving up the chain, processors with on-chip debugging resources will require external hardware support in the form of JTAG or another debug protocol.

What is significant is that these different processors all require different software designs and debug tools.

Embedded systems are typically extremely cost-sensitive

Several years ago, I had a VCR that died after the power fluctuated due to a windstorm; this was a rather poor power supply design. The VCR was very inexpensive, under $50. Out of curiosity, I took it apart and saved the main PC board to show my classes because it was a wonder of low-cost engineering. My guess is that the total manufacturing cost had to be under $20.

Yes, the unit was made in a low-wage foreign country, but the value engineering was beautiful. The main printed circuit board was single-sided. All the traces were on the bottom side of the board. Surface mount and through-hole parts were loaded on both sides of the board. Most of the parts on the top were jumper wires needed to cross over traces below.

I don't think the control processor cost more than $0.50 in high volume. I can't think of a better example than this VCR to demonstrate the cost sensitivity of embedded systems versus other computing devices.

Of course, with some exceptions such as NASA or the military, most products with microcontrollers are engineered for the lowest possible cost. So how does software fit into this equation? Simple. The software must be engineered to fit into the available code space to meet the cost and performance goals of the project.

Therefore, I would assert that our definition of what constitutes debugging should be broadened to include finding ways to improve the efficiency of the code in order to meet our design goals. The implication of this is that any code that may run but not meet the performance goals for the product is defective code.

Similarly, code that doesn't fit into the available code space (ROM, FLASH, EE prom, etc.) is defective, unless it can be shown that the specified feature set cannot be met with the available code space at the price point set for the manufacturing costs. Then we have a hardware problem as well.

Embedded systems have power constraints

About a year ago, I bought a highly regarded security system for my home. Totally wireless. All the remote sensors run on little lithium CR2016 button batteries. I have not replaced any of the batteries so far and the system is still going strong. What's interesting is that the base station is in constant communication with all the peripheral sensors and will send a message to the keypads if a sensor is not responding or needs a battery replacement.

Low power operation is a fact of life in today's embedded world. How does this impact software? The software must be architected and written for a low-power existence. The lower the clock speed, the lower the power consumption. Also, sleep mode designs must be used only to wake the processor when something needs to be done.

I was teaching an advanced embedded systems project class and one team of two students decided to build a battery-operated lock system—the kind you might find in a hotel room safe. Not a particularly challenging project, but I was curious to see how the team handled the power issues. Bottom line, they did not handle the power issues. They chose a microcontroller with low-power capabilities, but they never used these options, so their system ran for only about 15 min with a 9 V battery.

Embedded systems must operate under extreme environmental conditions

Embedded processors must go where the action is. They are required to operate under extreme temperature variations, humidity, vibration, and, in the case of spacecraft, the high radiation levels outside the Earth's atmosphere. Does this impact the process of software debugging? Probably not, unless you are trying to figure out why your Mars Rover is continually resetting itself.

Embedded systems have far fewer system resources than desktop systems

Suppose that you are writing an application for your PC. The PC architecture, whether its Windows, Mac, Android, or Linux, is a standard environment. For debugging purposes, you interact directly with the application via the resources available to you. The display and keyboard are your windows into the behavior of the code.

The vast majority of circuits containing embedded devices do not have keyboards or displays. It is also highly unlikely that they have communications ports that would allow you use a remote debugger with a debug kernel on the processor.

Embedded microprocessors often have dedicated debugging circuitry

On the plus side, almost all today's microcontrollers have some on-chip debug resources. I did an informal survey of the cheapest microcontrollers; even the $0.25 ones had debug cores. Most supported a JTAG-style debugger communicating with a host debugger; others have their own variants.

The Atmel AVR ATTINY13A-SSUR is unique because it needs only one wire to the RESET pin to communicate with the processor core.[e]

The advances in integrated circuit technology have made obsolete many of the former truths about the economics of IC fabrication. Today, transistors are basically free and there is no downside to placing robust debugging circuitry on the embedded microcontrollers, even though the debug core will never be used once the product has been designed, debugged, and shipped.

If an embedded system is using an operating system at all, it is most likely using a real-time operating system

If your embedded system uses an RTOS, it is highly likely that it is *not* a simple system and debugging it will be a challenge. It's no wonder that embedded systems require specialized tools and methods in order to be designed efficiently. Think of the various dimensions that must be balanced in order to achieve the final product. That's why understanding the use of these tools, even though they may be devilishly complicated to understand and require a learning curve you don't have time to climb, may be the only life preserver floating between you and a canceled project. RTOSs are complicated and can lead to subtle interactions between the hardware resources, the processor, and the applications under RTOS control.

Therefore, our software debugging objective is to make sure we understand how to use the tools both internal and external to the CPU core to find and fix bugs in the most expeditious way possible so we can ship quality products on schedule.

Yes, I know that sounds like every marketing pitch you've ever heard, but frankly, that's what we're trying to do. I use my students' experiences as examples of what not to do, but they have an advantage over embedded engineers in industry. For the student, every failure is a learning experience that has no real-world consequences (except for their grade in the class). However, if an experienced engineer makes a mistake that results in a defective product, you may be facing a career-defining moment.

Anyway, let's move on with a positive outlook about the future.

[e]I assume that you also need a ground connection, so that would make it a two-wire interface.

What are the best practices for debugging an embedded system?

We can divide this discussion into several categories:

- Avoiding the need to debug in the first place.
- Best practices for debugging software in general.
- Best practices for debugging embedded software in particular.

In the following sections, I will lean heavily on the writings of the experts. I'll give a brief overview of each "best practice." Then, I strongly recommend that you follow the links to the source of the reference and read the entire article.

Avoiding the need to debug in the first place

Much has been written about best practices in software development in general and embedded software in particular. Because whole books have been written on the subject, and software engineering is a part of every CS student's curriculum, I'll just touch on a few best practices I've found to be particularly relevant.

Partitioning

Ganssle [7] points out that one of the most fundamental insights of software engineering is to keep functions small, with about 50 lines of source code as a general maximum size. He discusses how large complex functions are difficult to understand and maintain.

The general idea is to compartmentalize a project into small chunks that can be worked on by small teams, generally working in isolation. Otherwise, all the issues associated with the communication between team members and between teams cause any project schedule to grow exponentially and to dramatically reduce the efficiency (measured in lines of code per engineer per month) by almost a factor of 10 as the project size grows from 1 to 100 in engineer-months.

Ganssle also describes another way to partition a design: throw transistors at it. Use multiple small microcontrollers, each doing one task, rather than a complex multitasking CPU running an RTOS with the associated issues.

Assertions

Beningo [8] and Murphy [9] advocate for the use of assertion macros in coding as a way to return an error message at runtime if the asserted condition is false. They point out that the ASSERT density can be the difference between a long debugging session and the discovery of a failed assumption at the moment that it first occurs. Murphy discusses the process for creating the ASSERT macro and placing it in your code.

Use lint

Lint is a static analysis tool that is helpful in analyzing C and C++ syntax to compensate for the C language's minimalistic approach to error checking in favor of speed and efficiency. The less runtime error detection in your code, the less software overhead there is to slow it down, and the smaller the software footprint.

According to Ward [10],

> *Lint is picky enough to satisfy even the most persnickety user. It identifies technically erroneous usage and it flags technically correct usage that is considered "poor style."...*
>
> *Lint even flags usages that are legal but a frequent source of inadvertent mistakes.*

Ward goes on to say that,

> *Lint produces a torrent of analytic information. It is wise to use it once early on to identify uninitialized variables and other blatant problems, and again during a code polishing stage.*

Another valuable feature of lint is that it will point to variables that are not used and code that illogically cannot be reached during program execution.

I think Lint is particularly useful for students who are interested in learning more about the nuances of the C language, but it can also be used to unwind a careless debugging session and restore code to working condition.

Gimpel Software[f] is one of the best-known companies building static analyzers. Their PC-Lint Plus is an industrial-strength tool that can analyze host-based and embedded software as well. PC-Lint Plus will also "check for violations of internal code guidelines as well as analyze compliance with industry standards such as MISRA."[f,g]

If you don't want to buy an industrial-strength product, you can find freeware versions of Lint on the web. AdLint[h] is an open source and freeware static code analyzer. It is available under the GNU general public license.

[f] www.gimpel.com.
[g] MISRA: Motor Industry Software Reliability Association, https://www.misra.org.uk/.
[h] http://adlint.sourceforge.net/.

One last comment about using Lint. Out of curiosity, I went back to the textbook I used when I taught an introductory C++ programming class. I saw no mention of Lint in the index, which reinforces the idea that teaching programming generally fails to also cover how to write good programs.

Take compiler warnings seriously

It's easy to ignore a compiler warning because your code still compiles. You can start the makefile and go to lunch. But compiler warnings are indications of potential bugs lurking in your code. They could be typos, such as using the = assignment operator when you mean == to test for equality.

Allain [11] points out that compiler warnings will point you to issues that might be difficult to find during testing. He uses the example of an uninitialized variable whose value can be different every time you test the code.

He goes on to say,

> *If you don't understand what a compiler warning means, it's probably best to trust that the compiler is telling you something valuable. I've had experiences where I was convinced that my code was correct and wasn't entirely sure what the compiler could be complaining about. But after investigating my compiler's complaints, I realized I had made a subtle mistake in a Boolean expression that would have been nearly impossible to hone in on when debugging–it was much easier to catch the mistake while the code was fresh in my mind code rather than hours or possibly days later when I might (with luck) have found the bug due to the mistake.*

Avoid software timing loops

I'm guilty of this one. Stewart [6] points to this as a bug waiting to happen. Perhaps not to you then and there, but whenever the processor, clock rate, or memory timing changes, the possibility exists that there will be a breakdown that can be hard to find.

Global variables

There was a very nice and well-meaning computer science lecturer on my campus who would always get really annoyed with me. This was because I told students in my Introduction to Embedded Systems class that there are times when a global variable is a reasonable solution. Her students who had taken my class would throw my statement back to her when she said that global variables are a Satanic invention. She was more correct than I was, but students failed to grasp the relative magnitudes of the arguments. One obvious place where a global variable is called for is when an interrupt service routine must exchange data with the main program. Because the ISR is running in its own context, there isn't a convenient way to exchange data. That's not to say it's impossible, just inconvenient.

Stewart points out that global variables should never be used to pass arguments. While it is convenient, he points out that it prevents the reuse of code.

Naming and style conventions

Both Ganssle and Stewart point out the need for coding and style conventions within an organization. For Stewart, a lack of style conventions, such as a consistent naming convention for variables, is the number one mistake for nonreal-time programming. Without a set of coding guidelines for naming functions and variables as well as the way code is represented, every programmer is speaking a different language and you'll need translators to understand each other's code.

The International Obfuscated C Code Contest (https://www.ioccc.org/) takes poor (or no) style conventions to their logical conclusion. The goals of the IOCCC, as stated on their web page, are:

To write the most obscure/obfuscated C program within the rules.

To show the importance of programming style in an ironic way.

To stress C compilers with unusual code.

To illustrate some of the subtleties of the C language.

To provide a safe forum for poor C code. ☺

Macho coders make it a point of pride to demonstrate how well they have mastered the precedence of operators and ignore parentheses that mere mortals need help to decipher. Parentheses provide guidance and eliminate the subtle bugs that would otherwise creep into your code.

Ganssle points out that if your organization has a 200-page manual on coding conventions, the manual will just gather dust. He advocates for a brief, easy-to-read booklet any engineer can pick up and readily understand the commonly agreed upon language conventions for the organization.

Code defensively

I found this lecture online [12] that I thought summed up very well the philosophy of defensive programming. The author defines defensive programming as a technique where you always assume the worst from all inputs. He identifies three rules of defensive programming:

> *Never assume anything. All input must be validated against a set of all legal inputs. Then determine the action you will take if the input is incorrect.*
>
> *Use coding standards (This one is familiar).*
>
> *Keep your code as simple as possible because complexity breeds bugs. Functions should be viewed as a contract between the user and the coder. The coder guarantees that the function will execute only a specific task and if it cannot do that task (divide by zero), it can notify the calling function that an error has occurred.*

Peopleware

Peopleware is the classic book on the sociological aspects of software project management by DeMarco and Lister [13]. It had a profound effect upon me when I read the first edition in 1987. I was an R&D project manager at the time, managing software and hardware engineers.

So much of the book talked about common sense issues that somehow get lost in the shuffle of working for a for-profit company. While I tried to point out best practices for the individual engineer, the authors take the wide view and look at the engineering environment that either sets up an engineer for success or failure. So, if you're a hiring manager and you want successful embedded software engineers, then you must provide them with an environment where they can be successful.

My number one favorite gripe from my days as an engineer and engineering manager is the noisy environment where my R&D Lab was located. We were seated in open cubicles with low walls, right next to the production area. While that may have been good for communication with the production engineers, it was horrible for thinking. To make matters worse, the PA system was constantly announcing phone calls and other administrivia for everyone to hear.

Engineers in my group routinely asked permission to work from home so that they could get their work done. I tried and failed to convince our management that we needed quiet spaces for engineers to work to concentrate without interruptions.

What else contributes to an overall poor engineering environment? Here's a short list:

(a) Unrealistic schedules: Create a schedule that you know you can't meet even under ideal conditions and then drive the engineers to track that schedule.

(b) No time for code reviews or testing: Engineers should have regular code inspections and be required to stress-test each other's code.
(c) No time for self-improvement: If every new project demands an engineer's energy with no time to unwind, read journals, or go to a conference, an engineer will get burned out and stale.
(d) Don't use bug reports as a punitive tool: Encourage engineers to find and fix bugs. Work with them to identify weaknesses in their techniques in order to help them grow.
(e) Do project postmortems: After a project is completed, have a trained facilitator sit down with the team and identify what went well and what could be improved. Feed the data back into the R&D lab processes to make sure mistakes aren't repeated. Ideally, the facilitator is someone from another division, group, or outside consulting organization not in the engineers' chain of command.
(f) Reward engineers who have an extensive sphere of influence: Identify not only the best engineers who solve the toughest design problems, but also those who help other engineers in the lab become better at their job.

Best practices for software in general

Whether you are debugging a complex algorithm, a hardware glitch, or a noise under the hood of your car, the process is pretty much the same. I'm reasonably confident that in our heart of hearts, we know the right way to do things. The key point is that even though we know what we should do, we often just don't do it.

There are lots of excuses, and many of them are pretty good excuses, as to why we ignore best practices and forge ahead with the shotgun approach, or the "Let's just try this and see what happens" approach. You're under a time crunch. But we're always under a time crunch. I spent all-nighters in the R&D Lab trying to get a product working so we could show it just like every other hardware engineer. I'm just as guilty as the next engineer.

For me, the clouds parted, and a shaft of light came down at about 3 a.m. on a Monday morning after spending the better part of the weekend in the lab. The thought popped into my head, "Right now, I'd rather be home in bed with my wife." At that point I decided to get serious about the best use of my time and try to minimize the last-minute heroics I had come to depend on for success.

This new approach extended beyond my debugging technique (or lack thereof). I got serious about creating and tracking my task lists and project schedule. I started to get other folks' opinions about design issues and to seek out informal and formal design reviews.

About the same time, I came upon Tom DeMarco's book, *Controlling Software Projects* [14]. I highly recommend this book, not because of the software management processes he describes, but because he discusses the real cost that leads to poor processes and wasted time.

I'm paraphrasing here, but he describes how some managers count divorce as a sign of toughness, like a gun fighter's notches on the pistol grip. DeMarco describes how unreal development schedules, lack of clarity in requirements, creeping features, etc., all contribute to engineer burnout.

Therefore, my crusade here is simply to make debugging embedded systems an engineering discipline with the same rigor we would apply to design engineering.

Develop a process

So many articles discuss this point and I teach it as well. There are any number of ways to basically describe the same process. My first lesson to my students is to ***write it down.*** Don't be tempted to dive into the code and start making changes in heroic attempts to do a quick fix.

When you first notice the bug, or anomaly, or whatever doesn't look kosher, write it down in your lab notebook, or however you record these things. Indicate the time of day and date, the conditions of your testing, what the system was doing when it failed to do what you expected.

Next, write down several hypotheses about what you think might be causing the problem. Next, talk to another engineer and bounce your ideas off him or her. Get their input and incorporate it into your set of possible causes. Then, if possible, systematically design a test case to isolate the cause of the defect.

There is another benefit to writing things down. You're forced to step back and reflect on possible errors as you review the code in your mind and to develop a deeper understanding of where the bugs might be lurking.

Beningo makes an interesting point here. He suggests rereading the data sheets and user manuals to make sure that registers are set up properly. Unfortunately, the key piece of information may be buried and finding the problem may take a bit of reading.

I remember one project for which we were building a software performance analyzer (SPA). The analyzer mostly worked, but it had a nasty habit of periodically (roughly every few weeks) producing an erroneous result. After many hours of debugging by many smart engineers, the error was traced to a typo in the manufacturer's description of the way a timer signaled when the count overflowed and rolled around to zero again.

I've watched students burn weeks of time trying to find bugs in their code and just digging a deeper and deeper hole even though they've been taught there is a right way to do this. Fortunately, the smart ones will quickly throw in the towel and ask for help.

However, "process" is much, much more than a process for debugging. *Process* is the core of software engineering. If you have fewer bugs in your code because your software was developed as part of a process, then the debugging process goes a lot quicker.

But even when you follow best practices as I've touched on earlier in this chapter, you're going to eventually have a defect in your code. Hopefully, you find it early enough in the process that schedules are not adversely impacted and, even more importantly, the bug is not discovered by a customer.

Generally accepted defect rates, taken from industry data, yield approximately eight defects per 1000 lines of source code, or just under 1%. This percentage is a measurement taken over all projects, so I would guess that code destined for mission-critical applications has a much lower defect rate. I became curious about defect rates in the best software in the world. I found this online discussion about the software on the space shuttle [15].

- The shuttle software consists of about 420,000 lines. The total bug count hovers around one. At one point around 1996, NASA engineers built 11 versions of the code with a total of 17 bugs. **Commercial programs of similar complexity would have thousands of bugs.**
- The software writing process was beyond meticulous. They needed an update to the GPS software (6300 lines of code). Before any work was done on the code, they wrote a 2500 page-long specification detailing the changes needed.
- Every change to the code was documented.
- Every bug was analyzed exhaustively.
- Everything the software group did was structured in processes, and every error a reason to examine and improve the processes even more.

Where do we go from here? Fortunately, enlightened CS faculty have begun to recognize the need to teach debugging as part of their undergraduate curriculum. Tatlock [16] included this statement in his lecture notes for CSE 331, *Software Design and Implementation*,

> *One of the painful parts of teaching a lab-based embedded systems course is that over and over I have to watch a team with a relatively simple bug in their code, but who is trying to fix it by repeatedly making random changes. Generally, they start with code that's pretty close to working and break it worse and worse. By the end of the lab they're frustrated, aren't any closer to finding the bug, and have made a complete mess of their code, forcing them to go back to the previous day or week's version.*
>
> *A typical Computer Science curriculum fails to teach debugging in any serious way. I'm not talking about teaching students to use debugging tools. Rather, we fail to teach the thing that's actually important: how to think about debugging. Part of the problem is that most CS programming assignments are small, self-contained, and not really very difficult. The other part of the problem is that debugging is not addressed explicitly. After noticing these problems I started to focus on teaching students how to debug during lab sessions and also made a lecture on debugging that I give each year....*

Here are some excerpts from Professor Tatlock's lecture on debugging.

The debugging process:

Step 1—find small, repeatable test cases that produce the failure

- May take effort, but helps identify the defect and gives you a regression test.
- Do not start step 2 until you have a simple repeatable test.

Step 2—narrow down location and proximate cause

- Loop: (a) Study the data, (b) hypothesize, and (c) experiment.
- Experiments often involve changing the code.
- Do not start step 3 until you understand the cause.

Step 3—fix the defect

- Is it a simple typo, or a design flaw?
- Does it occur elsewhere?

Step 4—add test case to regression suite

- Is this failure fixed? Are any other new failures introduced?

He illustrates this process in Fig. 3.2. Note the general nature of this diagram. He could be talking about debugging an emissions control system in an automobile engine. The process is fundamentally the same as software.

Automotive engineers will take the debugging process one step further. If you purchase the shop repair manual for your car, in each chapter such as engine, chassis, exterior, interior, etc., you'll

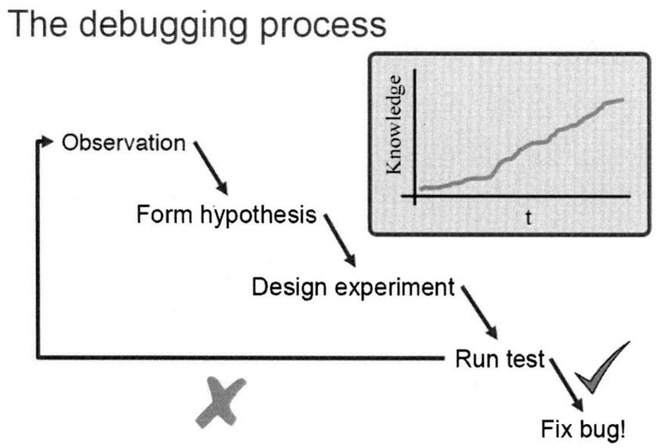

Fig. 3.2 Waterfall diagram highlighting the key steps in debugging. Courtesy of Professor Zachary Tatlock, University of Washington.

see a subsection titled *Troubleshooting.* It gives the shop mechanic a list of possible defects and provides a flow chart indicating how to isolate the defective component and replace it.

These are all excellent suggestions, but I've found that every seasoned engineer (those who have lasted through several rounds of layoffs) has their own debugging process that works for them. Like most others, I try the simple stuff first, and then, only when I'm stuck will I bring in the big guns. So, here's my process for debugging software. I'm sharing this only because it works for me.

1. Get out the lab notebook and start taking notes.
2. Try to make the bug repeatable.
3. Isolate the suspicious module. This may involve removing the module and testing it again in isolation, which means writing throwaway code, but that's the cost of doing business. If I can't isolate the problem to module, then go to step 4.
4. Change only one thing at a time and compare results. The shotgun approach is deadly and a black hole of time. Don't use it.
5. Divide and conquer. Use ***printf()*** function calls about half the way through the code and output the values of key variables. If everything looks good, then the problem is in the second part of the code. Repeat the process until I can zero-in on the region where the bug is occurring.
6. Ask for help. Show the code and demonstrate the bug to another engineer who I trust. I strongly believe that the best engineering comes through interaction. However, this means that:
   - **a.** When asked, I must be willing to look at their problems.
   - **b.** If someone helps me, I must give them appropriate credit.

People who are willing to help you will quickly disappear if you take personal credit for their assistance. There was one software engineer at HP who was my go-to person when I had a software bug that was eluding me.

7. If calling in the cavalry doesn't work, then it is time to start using tools. In order of preference:
   a. Debugger
   b. Logic analyzer
   c. In-circuit emulator

   We'll discuss these tools in a later chapter but, simply put, the emulator is a superset of a debugger and a logic analyzer in one integrated package. The logic analyzer and emulator are tools for debugging real-time systems. Because those are the systems I worked on, those were my tools.

I'd like to wrap up this discussion of best practices for general debugging with some observations:

- There is a lot great information on debugging techniques for the taking, but you need to be willing to invest the time to read and absorb it. As a professional, you have an obligation to stay current in your field and, if you don't, newly minted BSCS grads, up on the most current technology, are willing to do your job for less money than you earn.
- I've included a reading list of articles I've used as sources for this chapter for your use.
- There are many other specific techniques and processes you can adopt, and they are spelled out in the literature and in many of the articles in my reading list. I've only scratched the surface.

Best practices for debugging embedded software in particular

I've been thinking a lot about the differences between debugging embedded (or real-time) software and the general debugging issues that were discussed earlier in the chapter. The obvious differences are easy to enumerate:

- Must operate in real time.
- Debugging usually involves hardware interactions with software (HW/SW).
- Issues associated with a real-time operating system (RTOS).

However, what I think is the big separator is that embedded system failures are often nondeterministic. With external events occurring randomly and asynchronously, it can be nearly impossible to artificially recreate the exact sequence of events that led to the system failure. I purposely used the word "system" here because an apparent software failure can be traced to a hardware-software interaction issue, not necessarily a hardware failure.

Compare this with coding on a standard platform such as a PC, MAC, or Linux workstation. These are standard platforms with well-determined interfaces to the underlying hardware. The operating systems have been wrung out by tens of millions of users. If there is a software defect, then the process of finding and fixing the bug is straightforward.

This is the reason we need specialized tools that are tightly identified with the development of real-time systems—there is usually no other way to find an error. Therefore, in this discussion of best practices for debugging embedded systems, it is necessary to touch upon these specialized tools, even though we haven't yet discussed them.

I'm going to follow the same general flow model here that I followed in prior sections. I've culled the literature, I'll look at some of the most common bugs that crop up in embedded systems, and we'll look at their root cause and how to find them.

The only qualifier we need to mention here is that just about everything previously said also applies to debugging embedded code. Of course, embedded code adds its own set of headaches, but the process is the same up to a point:

- Write everything down.
- Change one variable at a time.
- Divide and conquer.
- Use assertions.[i]

Memory leaks

Memory leaks are associated mostly with systems that use dynamic memory allocation. You can have memory leaks in desktop systems as well as embedded systems, but there are several key differences. Embedded systems typically have much less RAM available for memory allocation than your desktop system; my new PC, for example, has 32 GB of RAM. Embedded systems are also expected to run for long periods of time without failure, and for a mission-critical application, a crash due to a memory leak is very dangerous.

[i]This may be a place where we begin to branch because performance-driven systems may not be able to output to the standard I/O output, stdout, during normal operation.

About 20 years ago, the world had the Y2K scare. People were worried that civilization as we know it would grind to a halt as computer system after computer system failed because they could not resolve a date change from 1999 to 2000. It is a tribute to the efforts of many, many people that the scare turned out to be a bust because so many systems required operating code patches or replacement.

I became involved with this issue in the summer of 1998 when an engineer in the electric power industry phoned me. They asked if I would be willing to consult on the Y2K problem. I told them, as far as I knew, there was no Y2K problem and I was about to hang up when he said his team was responsible for Y2K remediation at a coal-fired power plant. This plant used more than 500 embedded control systems and they didn't know which ones might fail if the real-time clocks in the devices erroneously changed the date.

Somewhat humbled, I agreed to help. They asked me to give a talk at a conference sponsored by the Electric Power Research Institute (EPRI) to the people from various industries tasked with dealing with Y2K remediation in their companies [17]. I later followed up the talk with an EPRI technical report [18].

What does this have to do with memory leaks? One of the other conference talks really hit home and explained the problem in a way I could understand. The speaker related how they approached the vendor of their data concentrator—a device that collected sensor data from remotely located transmitters, concentrated the data packets, and then periodically sent the packets to the control room.

The vendor was willing to help them fix the problem rather than try to sell them new concentrators, so the vendor patched the firmware and sent the power plant personnel new ROMs. On a Sunday night, the plant was shut down and the remediation team came in and replaced the firmware in each concentrator with patch code that supplied the date in a four-digit format, rather than two-digit, as it had been programmed to do many years earlier.

Nothing else was changed, except the size of the memory allocation call, malloc(), to make it two bytes bigger to accommodate a four-digit year. Every time a data packet needed to be sent, malloc() grabbed some free memory off the heap. Every concentrator was turned back on and passed its self-test. The remediation team went home to get some sleep. Early in the morning. The plant was turned back on to prepare for the Monday morning rush hour. Within 20 min. Every data concentrator failed, and the utility had to buy power from the grid until the remediation team could reassemble and replay the new firmware with the old firmware.

The plant worked fine with the old firmware reinstalled, but the utility was out several million dollars buying power from the grid.

A root cause analysis determined two problems. First, changing the size of the dynamic memory allocation meant that fewer packets could be processed before the concentrator ran out of allocatable memory. This would not have been so bad, but the manufacturer of the concentrator never put in an error handler for a malloc() error, which boiled down to the trap vector that was called to handle the error was a "jump to myself" instruction and, quite naturally, the systems all went down.

The question to ask about memory leaks is, "Does every memory allocation call have an associated memory free call (destroyer)?" If the system is running under an RTOS and a task is used to call malloc(), and then the pointer to the memory block is passed via a message queue to the task requesting memory, does that task deallocate the memory block? You need to look for that.

However, even without an RTOS, dynamic memory allocation is always a place to start looking for problems.

Memory fragmentation represents another dimension of problems with dynamic memory allocation. Fragmentation can arise when the size of the memory block being requested is variable. If every memory allocation requests the same size block, then one block is as good as any other block, and the only problem that can arise is running out of memory blocks to allocate.

Fragmentation happens when there is memory available on the heap to allocate, but it isn't the right size block, even though several noncontiguous blocks are available that have enough capacity to meet the allocation need. Because they are not contiguous, they can't be used.

According to Barr [19].

> *Fragmentation is similar to entropy: both increase over time. In a long running system (i.e., most every embedded system ever created), fragmentation may eventually cause some allocation requests to fail. And what then? How should your firmware handle the case of a failed heap allocation request?*

Barr suggests that one way to handle this is to limit yourself to one block size or, failing that, have multiple "heaps" with each one then designated for a single size block request. Then, write your own handler functions to preselect the proper heap based upon the size of the block being requested.

Fragmentation and memory leaks can be avoided by not using dynamic memory allocation. This is not always possible, but if you are tasked with designing a high-reliability system, then it might be worth the effort to investigate alternative system architectures that are not susceptible to this problem.

Java has memory management built into the language. Embedded programmers generally argue that having garbage collection[j] will make the code nondeterministic and will cause performance issues. This is a valid point, but I'm sure it is a wonderful source of ongoing religious wars between Java and C programmers.

My recommendation is to accept that if you need to use dynamic memory allocation, then you will need additional code to gracefully handle and report the system status when memory leaks or fragmentation occur. This is just good defensive programming.

Jitter

Jitter occurs through the interplay of the RTOS, the external environment, and incoming interrupts. Jitter is the variation in the time when a task can run and how long the task takes to run. If a low- or medium-priority task is constantly preempted by a higher-priority task or incoming interrupts, it may take longer to run than the designer ever thought by just analyzing the execution time of the code in isolation.

This jitter may lead to unacceptable performance or to occasional glitches that are devilishly difficult to find. Barr suggests that if jitter is an issue, then the solution is to raise the priority of the task or turn it into an interrupt service routine (ISR).

I wanted to include jitter in this discussion because I was personally involved in the development of two hardware-based tools that measured variability in code execution time. SPA was the code name for the HP B1487 software performance analyzer, a plug-in board for the HP 64700 emulation systems [20]. The analyzer was able to watch the unique address points of functions and other key elements of the system and time-stamp these functions so that time intervals could be measured, and the data gathered reduced by the postprocessing software and graphically displayed.

The weakness of this tool is that it depended on the operation of the processor being visible to the outside world. In other words,

[j] Garbage collection is the term used for automatic memory management performed by the language, although the term can mean any time that the unused memory objects are returned to the heap.

as caches became bigger, the ability of the tool to make accurate measurements was diminished.

Code Test was developed at Applied Microsystem Corporation (AMC) and later sold to Metrowerks, which was sold to Motorola Semiconductor Corporation, which was spun off to form Freescale, which merged with NXP. Code Test was lost somewhere in the transitions.

I was involved in Code Test when I joined AMC in 1996. Coincidentally, I was working on a similar technology at HP, code named Farside,[k] when I left the company. Both Farside and Code Test used a similar technology, called instrumented code.

The concept of instrumenting your code is not new. You do it every time you insert a printf() in the code to send a message about the state of the code at a particular point in the flow of the program. What distinguished the Farside and Code Test techniques was that the instrumentation process was automatically carried out by the tool.

User code was preprocessed to insert markers at the entry and exit points of functions. These markers were noncached writes to specific memory locations that could be detected by the tool as the code ran at full speed. When a marker was received, it was time-stamped and placed in a memory buffer for postprocessing. This way, the minimum, maximum, and average execution times could be measured for each function of interest.

Priority inversion

Priority inversion is my personal favorite RTOS bug. As I discussed in Chapter 1, the most famous priority inversion-induced bug occurred on the Mars Rover. As Barr points out:

> *The risk with priority inversion is that it can prevent the high-priority task in the set from meeting a real-time deadline. The need to meet deadlines often goes hand-in-hand with the choice of a preemptive RTOS. Depending on the end product, the missed deadline outcome might even be deadly for its users!*

RTOS vendors are aware of this bug and many of the commercially available RTOSs have priority inversion workarounds in their APIs.

[k]The choice of code name gave us many opportunities to include Gary Larson's "Far Side" cartoons in our internal presentations.

However, even though the fix is straightforward, this is the kind of bug that no amount of testing will uncover until it is in the field (Mars) and the failure occurs.

Stack overflow

This bug is interesting for reasons that have nothing to do with the bug itself. Through casual conversations with other embedded engineers, typically over beers at the end of the day at a conference, I discovered that this bug was a favorite question to ask during a technical job interview. So, if you are a student EE and beginning to look for a job as an embedded systems engineer, then this tip alone may be worth the price of the book.

The stack is a last-in, first-out data structure managed by the processor itself. The last item placed on a memory stack will be the first one retrieved. Stacks are typically located at the top of RAM and grow down toward lower memory, or the heap, where other variables are stored. Each time a function is called, a stack frame, or block of memory on the stack, is created that contains all the local variables needed by the function, the state of the processors' internal registers, data being passed into the function, and other information such as the return address after the function is exited.

Because functions can call other functions, including themselves, a stack can keep on growing until it grows so much it begins to trash other variables stored in the heap. This is stack overflow.

Barr points out that stack overflow affects embedded systems far more often than it affects desktop computers. The reasons are:

1. Embedded systems generally have far less memory than a PC.
2. RTOS tasks typically have one stack per task.
3. Interrupt handlers also use these same stacks.

Like a priority inversion, a stack overflow is difficult to find in testing. It may never come up in testing, and the effects of an overflow may not be seen until long after the overflow occurred because the trashed variable may not be needed for a long time after it was overwritten.

Now, here's the specific interview question:

How could you determine if your stack is big enough for your program's code (or some variant of this)?

The answer is to preload the stack region with a known pattern that is highly unlikely to be written by any same program. Next, run the program for various lengths of time and check to see where the highwater mark ends up. Then, allocate enough of an overhead to prevent the stack from spilling out of its boundary.

If you have an RTOS, then you can dedicate a task to monitor the stack to make sure that the highwater mark is never exceeded. The task should also be responsible for collecting information about the overflow and gracefully exiting the program, or some other recovery method, such as closing the offending task or generating a reset.

This also leads to a second interview question:

Is there a problem with using recursion as a programming technique, particularly with respect to mission-critical software?

And a follow-up question:

What other programming technique could you use instead of recursion?[1]

Wrapping-up

This chapter only scratches the surface of what is available in the literature regarding debugging software in general and embedded software in particular. The principle of Darwinian evolution generally eliminates software developers who can't debug their code or who produce buggy code that gets into the hands of the end users.

I've also noticed that technology has made buggy firmware more acceptable because ROM code is most likely stored in FLASH memory and can be routinely fixed in the field. I remember attending a talk at an Embedded Systems Conference on how to write firmware that is field-upgradable. At the same conference, I saw a soda machine that phones home and regularly reports on its status.

We see so many embedded systems that require field software updates to fix bugs that we don't give it much thought. However, with the interconnectivity of embedded devices now a fact of life, cybersecurity is a clear requirement. As hackers uncover vulnerabilities in our web-enabled embedded devices, they have a launch point for further mischief on our home networks. Perhaps we should give coders a passing grade because their software is scrutinized and attacked like never before.

Tools such as compilers, debuggers, and static analyzers are only useful if you know how to use them. I'm reminded of the parable of the woodsman who each day has to work harder and harder to cut the same amount of wood because he never stops to sharpen his axe.

[1]Answer #1, YES. Answer #2, A FOR loop.

Today, unintelligible user manuals are being replaced by YouTube videos and that's a good thing. I've done my own car repair work for a long time. In the past, I tried to decipher the factory repair manual or one of the second source repair manuals with indecipherable photos or diagrams. Not anymore. I just find the YouTube video of the repair I want to do. I had a recent student who became an excellent Labview programmer by watching YouTube tutorials. However, you still need to invest the time to learn the proper use of a tool, and perhaps even more significantly, the feature set of the tool.

We also discussed best practices for debugging and the importance of a disciplined approach and process.

I'll conclude this chapter with a reading list that includes the sources I've used as direct references and other background articles you may find interesting and useful.

Additional resources

- Books
  - Ann R. Ford and Toby J. Teorey, *Practical Debugging in C++*, Prentice Hall, ISBN: 0-13-065394-2, 2002.
  - David J. Agans, *Debugging*, Amacon, ISBN: 0-8144-7457-8, 2002.
  - David E. Simon, *An Embedded Software Primer, Addison-Wesley*, 1999, ISBN: 0-201-61569-X, Pgs. 283–327.
- Articles
  - Gokhan Tanyeri and Trish Messiter, *Debugging embedded systems*, Clarinox Technologies, Pty, Ltd., http://www.clarinox.com/resources/articles/.
  - Anindya Dutta and Tridib Roychowdhury, *Debugging software/firmware using trace function re-usable components*, www.embedded.com, June 1, 2009, https://www.embedded.com/design/prototyping-and-development/4008297/Debugging-software-firmware-using-trace-function-re-usable-components.
  - *EMBEDDED SYSTEM DEBUGGING*, http://www.romux.com/tutorials/embedded-system/embedded-system-debugging.
  - Ilias Alexopoulos, *How to debug embedded systems*, EDN, December 11, 2012, https://www.edn.com/design/test-and-measurement/4403185/How-to-debug-embedded-systems.
  - Robert Cravotta, *Shedding light on embedded debugging*, EDN, Vol. 53, No 8, 2008, pg 29.

- David LaVine, *Six debugging techniques for embedded system development*, April 2, 2015 https://www.controleng.com/articles/six-debugging-techniques-for-embedded-system-development/?utm_campaign=TURL-SixDebuggingTechniquesArticle&utm_source=Blog.
- Stan Schneider and Lori Fraleigh, *The ten secrets of embedded debugging*, embedded.com, September 15, 2004, https://www.embedded.com/design/prototyping-and-development/4025015/The-ten-secrets-of-embedded-debugging.
- Philip Koopman, *Avoiding the Top 43 Embedded Software Risks*, Embedded Systems Conference Silicon Valley, San Jose, May 2011.

References

[1] Arnold S. Berger, Embedded Systems Design, ISBN: 1-57820-073-3, CMPBooks, Lawrence, KS., pg. 12, 2002.
[2] Arnold S. Berger, Embedded Systems Design, ISBN: 1-57820-073-3, CMPBooks, Lawrence, KS pg XVIII, 2002.
[3] D. Gates, M. Baker, The inside story of MCAS: how Boeing's 737 MAX system gained power and lost safeguards, The Seattle Times (2019). June 22.
[4] https://news.vice.com/en_us/article/kzxq3y/self-driving-uber-killed-a-pedestrian-as-human-safety-driver-watched.
[5] https://web.archive.org/web/20041128024227/http://www.cs.umd.edu/class/spring2003/cmsc838p/Misc/therac.pdf.
[6] D.B. Stewart, Twenty-five most common mistakes with real-time software development, in: Embedded Systems Conference, Boston, Class ESC-401/421, 2006.
[7] Jack Ganssle, The Art of Designing Embedded Systems, Second ed., Newnes, an Imprint of Elsevier, Burlington, MA, ISBN-978-0-7506-8644-0, Pg.37, 2008.
[8] J. Beningo, 7 Tips for Debugging Embedded Software, EDN Network, August 4, https://www.edn.com/electronics-blogs/embedded-basics/4440071/7-Tips-for-debugging-embedded-software, 2015.
[9] N. Murphy, How and When to Use C's Assert( ) Macro, The Barr Group, March, 2001. https://barrgroup.com/Embedded-Systems/How-To/Use-Assert-Macro.
[10] R. Ward, Debugging C, Que Corporation, Carmel, IN, 1990, pp. 62–66. ISBN: 0-88022-261-1.
[11] A. Allain, Why Compiler Warnings Are your Friends, Cprogramming.com. https://www.cprogramming.com/tutorial/compiler_warnings.html, 2019.
[12] A. Denault, Comp-206: Introduction to Software Systems, Lecture 18: Defensive Programming, Computer Science, McGill University, Montreal, 2006. https://www.cs.mcgill.ca/~adenau/teaching/cs206/lecture18.pdf.
[13] T. Demarco, T. Lister, Peopleware: Productive Projects and Teams, third ed., Addison Wesley and Dorset House, 2013. ISBN: 0-321-93411-3.
[14] T. DeMarco, Controlling Software Projects: Management, Measurement, and Estimates, Prentice Hall, 1986. ISBN: 0-13-171711-1.
[15] https://space.stackexchange.com/questions/9260/how-often-if-ever-was-software-updated-in-the-shuttle-orbiter.

[16] Z. Tatlock, From Lecture 15, Debugging, CSE 331, Software Design and Implementation, Winter Quarter, University of Washington, Department of Computer Science and Engineering, Seattle, WA, 2017. https://courses.cs.washington.edu/courses/cse331/17wi/lec15/lec15-debugging-4up.pdf.
[17] A.S. Berger, A brief introduction to embedded systems with a focus on Y2K issues, in: Presented at the Electric Power Research Institute Workshop on the Year 2000 Problem in Embedded Systems, August 24–27, San Diego, CA, 1998.
[18] A.S. Berger, A Primer on Embedded Systems with a Focus on Year 2000 (Y2K) Issues, Electric Power Research Institute Report #TR-111189, (August 1998).
[19] M. Barr, Top 10 Causes of Nasty Embedded Software Bugs, The Barr Group, May, 2016. https://barrgroup.com/Embedded-Systems/How-To/Top-Ten-Nasty-Firmware-Bugs.
[20] A.J. Blasciak, D.L. Neuder, A.S. Berger, Software performance analysis of real-time embedded systems, HP Journal 44 (2) (1993) 107–115.

4

Best practices for debugging embedded hardware

Introduction

I feel more comfortable writing this chapter because I was a hardware designer at HP before I was promoted to management and became a generalist.[a] I also teach microprocessor design at the University of Washington Bothell and the senior level students in my class must complete a design project in the class, complete with a printed circuit board (PCB) that they have to design. For most of the students (except those whose hobby is electronics), this is their first foray into PCB design.

Later, they must complete a more substantial project as part of their Capstone Experience. All accredited[b] EE programs are required to have a Capstone project as a required element in their degree programs. Our Capstone model has the students function as a small consulting engineering company [1]. This exposes our students to the entire product design lifecycle, including hardware turn-on, debugging, and validation testing.

I also manage the Capstone program for our department and have personally been the faculty advisor to many teams. I've probably observed more wrong ways to debug hardware that you can imagine. The inability of a student to logically analyze a problem, form a hypothesis of the cause, then test that hypothesis causes them to waste valuable time and destroy parts and printed circuit boards in the process. Observing these students was my primary motivation for deciding to write this book.

As in the previous chapter, this chapter will focus on best practices. Once again, I've culled the literature to find tips and tricks as well as application notes from the companies that provide the tools of the trade for the hardware designer.

[a]There's a Dilbert cartoon about this.
[b]www.abet.org.

Debugging Embedded and Real-Time Systems. https://doi.org/10.1016/B978-0-12-817811-9.00004-1

The debug process for hardware

I want to kick off this chapter by relating a story that was told to me during the years I served on the Project Management Council in Corporate Engineering at Hewlett-Packard. I also told this story in an earlier book [2], but its lesson is still relevant today.

About 20 years ago, the part of HP that is now Keysight® was rapidly moving towards instrument designs based upon embedded microprocessor. HP found itself with an oversupply of hardware designers and a shortage of software designers. So, being a rather enlightened company, HP decided to send willing hardware engineers off to software boot camp and retrain them in software design. The classes were quite rigorous and lasted about 3 months, after which time one of the retreads returned to his HP division and started his new career as a software developer.

This "retread engineer" became a legend. His software was absolutely bullet-proof. He never had any defects reported against the code he wrote. After several years he was interviewed by an internal project team, chartered with finding and disseminating the best practices in the company in the area of software quality. They asked him lots of questions, but the moment of truth came when he was bluntly asked why he didn't have any defects in his code. His answer was quite straight-forward, "I didn't know that I was allowed to have defects in my code."

In hindsight, this is just basic Engineering Management 101. While he may have been retrained in software methods, his value system was based on the hardware designer viewpoint that defects must be avoided at all costs because of the severity of penalty if a defect is found. A defect may render the entire design worthless, forcing a complete hardware redesign cycle taking many months and costing hundreds of thousands of dollars. Since no one bothered to tell him that it's OK to have bugs in his code, he made certain that his code was bug-free.

My point here should be obvious. Hardware is unforgiving and it is potentially very expensive to find and fix defects. We've made great strides in technology since then, particularly in the area of FPGAs where reprogrammability has turned much of hardware design into software design. However, you can't do everything in an FPGA. At some point, the signal must exit the FPGA and interact with the outside world or other software. Sometimes it interacts too strongly and then a perfectly performing system can't be shipped because its RF emissions are too great to allow it to be sold as a commercial product.

Design reviews

My number one recommendation for best practices in the category of "the easiest defects to find are the ones that aren't there" is to have other engineers review your work in the hope of finding problems before they become embedded in an ASIC or PC board. I teach my students how to do a design review in the faint hope that they will take my advice, but this is like trying to spit into the wind. The typical response is, "We didn't have time to do a design review."

The alternative behavior is for a student to come into my office and ask me to do a quick design review because they need to order their PCB in the next hour or so. Even worse, they show up with a PCB layout drawing and no schematic diagram of the circuit. They never bothered to do a schematic design.

A sane organization realizes the value of design reviews and makes certain that all project schedules contain time for engineers to have design reviews of their hardware and to have the time to serve on design reviews as examiners for other engineers.

We all know that the later into a project a defect is discovered, the more costly it is to fix the defect. This is true for software and for hardware, but more and more software (and FPGA-based hardware) can be repaired in the field with software updates, so perhaps this general rule is not as universally true today as it once was, but not every embedded system lends itself to field upgradability. Still, as a manager, would you prefer to have your engineers designing new products or repairing products that are out in the field generating customer complaints?

My very first hardware design review was an extemporaneous meeting between me and David Packard, the "P" in HP. I was fortunate to have started working at the Colorado Springs Division of Hewlett-Packard when both William Hewlett and David Packard were both alive and very much in charge of the company. They were at our division for the annual division review and "Dave" was wandering around the R&D lab talking with the engineers.[c]

I was working on a schematic design, not paying much attention to what was going on around me when a shadow suddenly appeared over my desk. I looked up to find that David Packard was standing over me looking at the schematic. He pointed to a part of the circuit and said, "That won't work." I said, "Yeah, it will." That was his way of introducing himself.

Several years later I was at HP corporate headquarters in Palo Alto and I happened to be walking down a hallway and I ran into

[c]In HP culture, this was referred to as "management by wandering around" or MBWA.

him. I said hello and introduced myself, and I reminded him of our first design review meeting. He smiled, but I don't think he remembered the meeting with the same clarity that I did, and still do.

Anyway...

What are the steps for a good process for a design review? A note here that the purpose is to uncover flaws and weaknesses, not rubber stamp a design so that a box could be checked off saying that you held a design review. A design review is labor-intensive, takes time and is very expensive in terms of lost productivity. When you take an engineer away from their own design work, there is a double whammy because the actual hourly cost of an engineer, including salary, benefits, equipment, etc., and on top of that, the lost productivity cost to their primary project becomes the cost of doing the design. Therefore, for a design review to have a positive return on investment (ROI), it needs to be effective and taken seriously by everyone involved in the process.

Step 1: About a week before the review, identify three or four other engineers who will take part in the review. Ideally, one of the engineers will be the moderator/recording secretary and the others will do the actual review of your design.

Serving on a review panel is like jury duty. You can get out of it once or twice with a good excuse, but you eventually must take part.

Step 2: Circulate all the pertinent design documents at least 3–4 days before the review. This would include schematics, data sheets, and ABEL, CUPL, VHDL, or Verilog code for the programmable parts as well as timing calculations, simulation results, data sheets, or app notes. In short, any of your design documentation should be part of the review material.

Step 3: To start the review, the engineer gives a brief overview of the design and covers the pertinent product requirement specifications. This might cover the choice of microprocessor or microcontroller being used, the amount of memory needed, clock speeds, and so forth.

Step 4: During the review, the moderator will lead the review process and the reviewers will bring up any design issues that need further attention. The purpose of the review is to identify issues, not to solve them. However, engineers being engineers, getting into a problem-solving mode is almost inevitable.

The designer being reviewed may offer some explanations but should not attempt to defend the design in the review. This also brings up another issue. The reviewers should comment on the design, not on the designer. Saying something like, "How could you make such a bonehead mistake?" just puts the designer on the defensive and the effectiveness of the review will quickly unravel.

The moderator notes each issue as it comes up, and whether it needs further follow-up by the designer. If the designer does not agree with the comment, defending the choice should be reserved for a later date.

Step 5: At the conclusion of the review, the moderator reiterates each issue that requires further attention and the meeting adjourns.

Step 6: The moderator then writes a summary of the meeting contents and the issues to be addressed. A copy of this report is sent to the lab manager and all the participants.

Step 7: The engineer whose design was reviewed then addresses each issue and does a final write-up report of how each issue was addressed. Sometimes, a follow-up review is scheduled if the number of issues was significant and the design was substantially changed.

Step 8: Once everyone signs off on the review, the design can move to PC board layout.

I can remember the details of my first HP design review like it was yesterday. I was very nervous because I was going to be reviewed by some of the top guns in the lab and I was the new guy. My design was a controller board for a 16-channel oscilloscope probe multiplexer with a 1 GHz plus bandwidth (HP 54300A).

During the review, one of the reviewers picked up on some high value resistors that were connecting the 8 data lines to the +5V power rail and to ground, seemingly at random. I told him that was for testing and the release version would not have the resistors loaded. The purpose was to force the data lines to always supply the processor with an NOP (no operation) instruction so I could probe the board and look at timing margins and signal integrity. He told me, “That’s a cool idea. I should try that.” I was very relieved.

There is an epilogue to this event. Years later, after I had left HP, I ran into one of my colleagues from the R&D lab at an Embedded Systems Conference. We were reminiscing about old times and he mentioned that my probe multiplexer had a very high failure rate, which I couldn’t understand. The problem turned out to be the high-frequency transmission line switches that I had chosen to use. According to the manufacturer, they were rated for several million cycles, which should have been more than enough in service.

However, these switches were originally designed to be used in missiles where they had to be reliable for a few hundred switches, and then they disappeared in the ensuing explosion. The manufacturer had never tested the switches to failure, and the failure rate quoted was only a guess. The design review would not catch this defect, but a more thorough investigation of the switch data, or the manufacturer of the switch, might have raised a red flag.

Bob Pease [3] was a legendary analog circuit designer at National Semiconductor (now part of Texas Instruments). In his classic book on troubleshooting analog circuits, he describes how he also conducts informal design reviews.

> *At National Semiconductor, we usually submit a newly designed circuit layout to a review by our peers. I invite everybody to try to win a Beverage of Their Choice by catching a real mistake in my circuit. What we really call this is a "Beercheck." It's fun because if I give away a few pitchers of brew, I get some of my dumb mistakes corrected. Mistakes that I myself might not have found until a much-later, more-painful, and more expensive stage. Furthermore, we all get some education. And, you can never predict who will find the picky little errors or the occasional real killer mistake. All technicians and engineers are invited.*

Test plan

The primary audience for this section is the soon-to-graduate EE student. Assuming that your resume is sufficiently compelling, a company will phone screen you and then if you pass that hurdle, invite you in for an in-depth interview. Along with the invariable technical interview, you'll probably interview with several of the managers. Their job is to assess your "other skills." This would normally include things like professionalism, communications, match to company culture, and maturity level.

Your job is to convince the interviewers that you are the best candidate for the job. Along with your technical ability, you need to convince them that you are ready to step into the maelstrom and be productive from day 1. They will invariably ask you about projects you've done and from speaking with our alumni,[d] we've learned that their Capstone Experience was one of the key factors in their getting hired. What impressed the interviewers was the "soft skills" that the students learned as part of the Capstone project. Skills such as tracking schedules, holding team status meetings, documentation, and having a test plan. The test plan is a real winner with interviewers.

The test plan is the roadmap that you will use to go from a raw PCB to a functioning prototype. Just like a pilot's checklist, it is your step-by-step guide to turn on and debug so that you do not miss anything, or worse, waste time and destroy your hardware.

[d] It is part of our accreditation process to invite graduates back for focus groups 3–5 years after graduation.

We give every new Capstone team an official lab notebook, just like the ones I had as an engineer. They are instructed to create the test plan and then document their work in the lab notebook. This is an unbelievable hardship for the generation of students who have a smart phone surgically attached to their hand, but they humor me because I have the power of the grade to hold over them.

To introduce the topic to the students, I pose this question.

Suppose that you've just gotten your raw PCB back from the board manufacturer. It's gorgeous. It is the first PCB that you've ever designed. What is the first test that you perform on this board?

After the blank stare goes away, I get a variety of answers, but rarely the correct one.[e]

Here is the format that I suggest for the test plan. Your mileage may vary.

1. Raw Board
   - a. Visually inspect raw board and compare with design
     Match/Problem?__________________
   - b. With ohmmeter, measure resistance between Vcc and ground
     Expected: Open circuit, measured____________
2. Loaded board
   - a. Are all parts properly located and aligned?
   - b. Are all socketed IC pins in the sockets and are the ICs oriented correctly?
   - c. Are all the leads soldered to the board?
   - d. Are there any solder bridges between pins?
   - e. Is there any residual solder flux on the board?
   - f. Are all the components seated properly?
3. Board turn on (no external inputs)
   - a. Measured resistance between Vcc and ground__________________
   - b. Is it a short circuit? Yes/No________
4. Apply power with current limit turned on or a series resistor in the power rail
   - a. Any odors? Yes/No__________
   - b. Any heat? Yes/No___________

   Note: This test implies that you've made a worst-case power consumption calculation at idle, so that if the board exceeds this during turn on you know something is amiss before one of the traces turns into a fuse.

[e] Answer: Check that power and ground are not shorted together.

5. Measure supply voltages
   a. Is Vcc correct on all inputs? Yes/No_______

And the list goes on. At each step, you are gaining confidence that your design is correct.

And so forth. While this may seem a waste of time, it is worth its weight in time saved and it definitely impresses your interviewers.

The test plan really becomes valuable when something doesn't measure what it should. Here's a simple example. Suppose you are designing a precision rectifier circuit. You apply a 1 kHz sine wave and observe on the oscilloscope that during the negative transition of the input, the rectified sinewave output, is missing.

An experienced designer would head straight for the rectifying diodes, realizing that one is probably soldered in backward or there is an error in the PCB connection. Maybe the silkscreen layer on the board is wrong. Whatever.

The best practice that I teach my students is to stop and write down in their lab notebook what they expected to see and what they observed. Because our oscilloscopes are connected to the network, I encourage the students to print out the display trace so that they can add it to the notebook.

Step #2: I tell them to think about how the circuit should work (theory of operation) and based upon their understanding of the circuit, write down several possible culprits. This will force them to reinforce their knowledge of the circuit behavior before they attack the board.

This is particularly true when a student grabs an example circuit off the web and tries to use it without really understanding how it works. If necessary, reread the data sheet(s) to make sure that you haven't missed something.

Step #3: Probe the board with the oscilloscope and note the waveforms and DC operating points at the circuit nodes.

Step #4: Run their design in a simulator, such as Spice, LTspice, or Multisim and try to recreate the fault by testing their hypothesis in simulation. Once they can demonstrate that the simulator will produce the same output as their circuit, then they may proceed to investigate the suspected part.

If you are an experienced hardware designer, at this point you are probably rolling your eyes. I know what you're thinking. But...

There are many years between you and that new engineer. Let's give the newbie every opportunity to be successful.

Design for testability

Way back in the Dark Ages, I wrote an article [4] about designing embedded hardware with an eye to tools that will be needed downstream to debug the system. Because I was involved in the design and manufacture of in-circuit microprocessor emulators (ICE), this was my primary focus in the article.

It was basically simple stuff, such as avoiding the positioning of the microprocessor in such a way that connecting an emulator or logic analyzer is impossible. Whenever you look at a marketing brochure, you see a single board sitting conveniently on an immaculately clean desktop with the instrument of interest prominently displaying an exciting waveform.

In reality, the real board is going to be crammed into a card cage with barely enough space for airflow. If you do need access to the processor for debugging, it will likely be a JTAG connector that you need to plug into. It won't do you any good if the connector is located on the side by the backplane socket or guide rails, and not the side by the board ejectors.

Along the same lines, design the first board as a prototype board, not as your final board. Give yourself a fighting chance to find problems.

Although Pease's book was dedicated to analog circuitry and discussed a fair amount about IC design bugs, there is a lot of practical information in it that is still useful today. In a section titled "Make Murphy's Law Work for You," he describes how he often allocates some extra space in certain parts of his PC board because he isn't 100% sure that it will work, so he's leaving room to make modifications in the next revision of the board.

Another really interesting tip I learned from reading his book is something that only an analog designer would think about but digital designers tend to ignore, which is signal fidelity. Pease was describing looking at a pulse train that had a lot of ringing in it. Because he was concerned about the fidelity of the pulses, he investigated it and found that the 6″ ground lead from his oscilloscope probe was producing the ringing. He fixed the problem by adding small ground pads near critical signal nodes that he wanted to probe.

Thus, using a probe with a ground ring close to the tip of the probe, as shown below in Fig. 4.1, he was able to eliminate the ringing caused by the inductance of the 6″ long ground wire clip.

I like to add through-hole pads in critical traces. The hole is sized so I can solder a small post to it and attach a probe. Also, it is almost absolutely required that you add ground pins and even Vcc pins, just in case you need to power a logic probe to examine the circuit.

Have a process

Here's where hardware and software part company. Software can easily be checked by a recompile and download cycle, assuming that you don't have an overnight software build process to deal

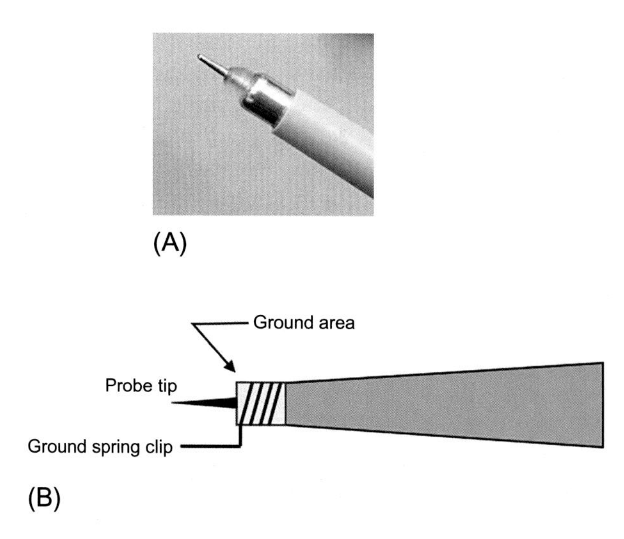

Fig. 4.1 Probe tip with ground area (left). Schematic drawing of clip with ground spring attached.

with. One of our claims with the HP 64000 in-circuit emulator system was that you could find a bug, fix it, and download a new software image in under 1 min. That's great marketing fluff, but it isn't a process either.

Putting that aside for a moment, we don't have the same luxury with hardware because hardware is physical and not ethereal. Before we start swapping components, cutting traces, or hanging filter capacitors on the circuit, we should have a high confidence level that the fix will work.

I think I discussed this point in every one of the preceding chapters, so hopefully it will stick. Like software, there are best practices for designing the hardware that you should follow in order to minimize the likelihood of introducing a defect into your design, or at least minimizing its severity.

Unless you're designing an FPGA, your objective is not to introduce a defect, just like the HP engineer in the example I cited earlier in the chapter. However, your process should include in your process plan the reality that you will likely have to do one or more circuit redesigns before the hardware is ready for prime time.

Also, because this is a design that will need to be integrated with software, we need to consider the very likely scenario that the hardware will check out just fine until the actual software is introduced, and then, and only then, will the bugs come to light. So, the process usually requires that there be an iterative aspect to the hardware debug phase.

Embedded system tool suppliers have been discussing this process for years and I have been as guilty as the next marketing drone because I spent a total of 19 years working for tool companies.

Here's the classic hardware/software (HW/SW) integration loop figure:

This flowchart describes the classic HW/SW integration process. Many people have argued that this chart is either misleading or it encourages bad practices because it is implying that the hardware and software are isolated from each other until the very moment when they are brought together, late in the product design cycle, and then integration can begin.

While Fig. 4.2 is conceptually easy to understand, it is considered to be rather backward. In fact, various experts would argue that HW/SW integration, test, and debug is an ongoing process and bringing the final versions together should be an anticlimax rather than a cataclysmic event.

There are lots of reasons why the HW/SW integration process is a major time suck to the product development schedule. However, assuming that the hardware works as the designer intends it to work, and the driver software works as the firmware designer intends it to work, then what's the problem? I would posit that a communications failure between the HW and SW developers accounts for the vast majority of the problems with the integration phase of the project.

Here's a simple example. The hardware designer ignored the Endianness of the system because all that needed to be done was to connect the peripheral devices to the address and data buses. Murphy's Law dictates that whatever the Endianness that the firmware developer assumes, it will be the opposite. Sometimes it is an easy fix because the processor has a register that enables the Endianness to be software-controlled. In the ARM 8 Cortex processor, the Endianness of the data can be set via a software register, but instruction accesses are always Little-Endian.

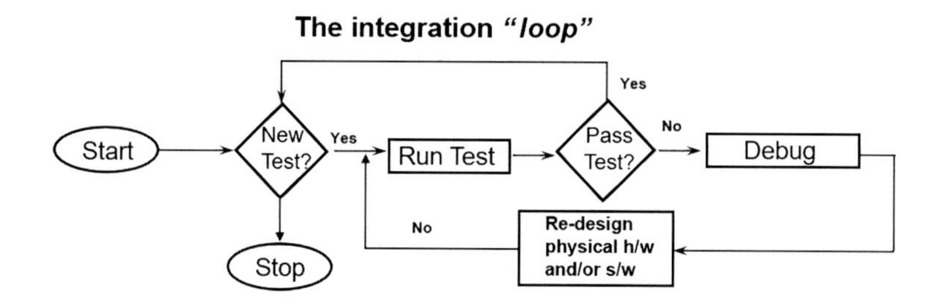

Fig. 4.2 Hardware/software integration loop.

Certain compilers such as GCC will allow you to set an Endianness switch, so the bug fix might only require a recompilation of the code. Still, a bug is a bug.

If a process was in place to address this possibility, then the internal formal specification for the product would spell out the Endianness of the system and software developers could build hardware simulation code that takes this into account. An important process requirement that could minimize the need for the HW/SW integration phase is to agree early on that the software team and the hardware team come together and create a detailed interface specification so that the software team could create a testing scaffold for their code that would provide a correct interface to the hardware under development.

Many RTOSs, such as Integrity from Green Hills Software,[f] have extensive support for hardware virtualization so that the software team can move forward with code development and integration with the RTOS while hardware is still under development. So, there really isn't a good excuse for not modeling the hardware so that incremental integration can take place.

David Agans wrote a very readable book on debugging [5]. He suggests a process that you can get in the form of a poster. Fig. 4.3 is a reproduction of this poster. I suggest that you put it up in your lab to remind yourself how to proceed, just in case you forget.

This is a wonderful little book and an easy read. You can easily finish it at a long session at Starbucks. In granting me permission to reproduce the poster, the author asked that I recommend the book, and I do recommend it. It has lots of practical suggestions and real examples from the author's embedded systems design experience.

Of the nine rules, the one that resonates for me is the last one, *If you didn't fix it, it ain't fixed."* Because I minored in English, I would say, "it isn't fixed," but the sentiment is the same.

It might be really tempting to ship a product with a bug that you've seen but can't seem to isolate. It's infrequent, so what's the harm? But you know it's there, it's lurking.

You might try to convince yourself that it was a glitch, or a hiccup, and it isn't "really" a bug, but in your heart, you know it's there, so you might as well find it.

However, suppose that management needs the product to ship and you are feeling the pressure to release it to manufacturing knowing that there is a defect in the product. How do you respond? This dilemma has been much studied, with the Boeing

[f]www.ghs.com.

Fig. 4.3 A debugging process. Reprinted from D.J. Agans, Debugging, the 9 Indispensable Rules for Finding the Most Elusive Software and Hardware Problems, AMACOM, 2002, ISBN:0-8144-7457-8, with permission; courtesy of the author.

737 MAX incident as only the latest to make the news. The IEEE website has a page on the Code of Ethics[g] that should provide you with some guidance:

> *to hold paramount the safety, health, and welfare of the public, to strive to comply with ethical design and sustainable development practices, and to disclose promptly factors that might endanger the public or the environment…*

My first job at the Colorado Springs Division of Hewlett-Packard[h] was as a CRT[i] designer. I was responsible for the CRT in the HP 1727A storage oscilloscope. Today, any modern oscilloscope is a storage scope because the waveform acquisition system is all digital. In the HP 1727A, an electron beam "wrote" the waveform on a dielectric mesh. Writing involved knocking electrons off the mesh, creating local areas of positive charge wherever the beam struck the mesh.

A second low-energy electron beam then flooded the mesh and any of those electrons that came close to the positively charged regions were able to get through the mesh and then were accelerated with a 25kV potential to the phosphor-coated screen, producing the stored image of the waveform. Anyway…

The scope had a bug. This is strictly an analog design. No processor in it, but it still had a bug. The bug was infrequent, and we could never find it. The bug was in the trigger circuitry that would cause the electron beam to fire once and record the waveform. Occasionally, for no apparent reason, the oscilloscope would trigger, but apparently the trigger signal did not come from an input source.

Now, the whole point of buying this $20K+instrument was to be able to capture elusive signals. It rather defeats the purpose if the instrument randomly fires and the customer never sees the real signal they were hoping to find.

For some reason, the engineers began to suspect that it was "microphonics," or mechanical jarring, that was causing the scope to misfire. So, we started to hit it with rubber mallets and drop it. Anything we could think of to reliably generate the fault, we tried, and we couldn't reliably recreate the failure. I think they eventually found it to be a high-voltage arc that found its way to the trigger board, and the problem was resolved.

To their credit, they did not ship any of these instruments until the problem was solved.

[g]https://www.ieee.org/about/corporate/governance/p7-8.html.
[h]Now Keysight Technologies, Inc.
[i]That's a cathode ray tube.

Know your tools

My involvement in embedded systems has always been on the debugging tool side of the industry, so I am understandably focused on this dimension of the process. One of the biggest issues that I saw and experienced many times over was the inability of our customer to understand how to properly use our tools for maximum advantage.

I've seen it with really senior engineers and with my students. Of course, we the tool vendors have to share some of the blame because we're the ones who created the tool and then did not provide adequate documentation to make it easy to understand how to use the tool to its fullest extent.

In the introduction to the book, I mentioned Hansen's Law. However, because most people never read the introduction, let me summarize it again. John Hansen was a brilliant HP engineer whom I had the privilege to work with in the Logic Systems Division in Colorado Springs. He said,

> *If a customer doesn't know how to use a feature, the feature doesn't exist.*

It's a very simple yet extremely insightful statement about designing complex products and the need to be able to simply convey their usefulness to an end user.

In his seminal book, *Crossing the Chasm* [6], Geoffrey Moore looks at the marketing of high-technology products. Moore identifies a fallacy in the traditional way that high-technology markets are modeled. Consider the traditional life cycle model for the adoption of a new product in the marketplace, shown in Fig. 4.4A. We see each segment of the market occupying a portion of the area under the bell curve. The area in their segment represents the potential sales volume for that market. I think we can easily identify the characteristics of each segment.

However, for the successful marketing and sales of new technology-based products, Moore argues that this model is wrong. He argues that there is a fundamental gap, or chasm, that exists between the early adopters and the early majority. Referring to Fig. 4.4B, we see that the segments comprised of the early and late majorities are the bulk of the market. Therefore, while initial sales to the "techies" might be very gratifying, those sales can't sustain a successful product for very long.

Moore says that the early majority are the gatekeepers for the rest of the market. If they embrace the product, then the product can continue to grow in sales and market impact. If they reject it, then it will die.

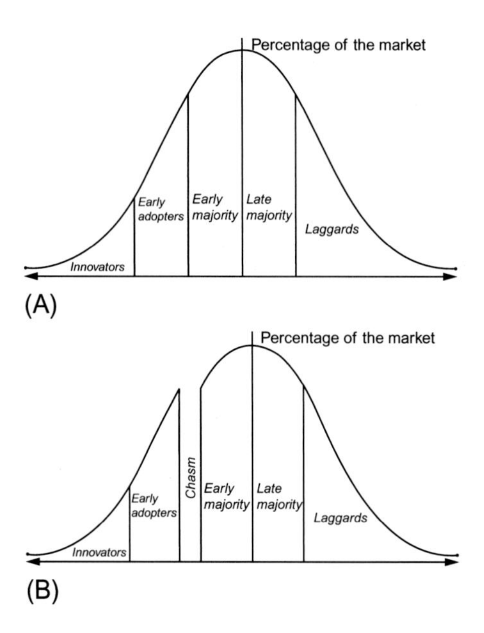

Fig. 4.4 Continuous product adoption life cycle. Part "A" is on the left and represents the traditional viewpoint. Moore argues that a discontinuous model, part "B" on the right, is the appropriate model for high-technology products.

In order to be accepted by the early majority, there are several key factors that must be in play, but I'll just zero in on two factors that I think are germane to the point I'm trying to make.

- The early majority tends to seek validation of a product's value by seeking the recommendation of other like-minded members of the early majority whose judgment they trust.
- There must be a "complete solution" available for the product.

The second bullet is the one that is relevant here.

As developers of new technology, we are constantly coming out with newer and better solutions to meet the needs of our customers, who are themselves developing new and innovative products. These customers, the early majority, don't have the time or desire to put up with the glaring omissions of a new product's support infrastructure. Manuals with errors, lack of technical support, training are all unacceptable show-stoppers for the early majority.

So, what's this got to do with "know your tools?" We, the R&D engineers, can provide all the features in the world to make our in-circuit emulator or logic analyzer more compelling, but if a customer can't take advantage of the feature set because it is too difficult to learn, or they don't have the time to learn, then the tool is not what they need.

Yes, the tool manufacturer needs to bring every technology transfer best practices to bear to make the features of their products accessible and easy to understand. In their defense, I will say that the tools today are much better than when I was working in the field. This is primarily due to the additional processing power built into the tools and the amount of memory that even the most modest of instruments carries within. Rather than trying to find it in the manual, I can press a context-sensitive help button and see the manual page that I need to access.

But… the onus is still on me to devote enough time to learn the tool. If I'm always too busy to learn how to take advantage of my tools, then I have no one to blame but me for not surviving the next round of layoffs. Just remember the parable about the wood cutter who was always too busy chopping down trees to sharpen his axe and then couldn't understand why he was not able to cut as much wood as he needed to.

Understanding your tools extends beyond a knowledge of how to use the feature set. It also includes an understanding of how the tool interacts with the environment that it is being used in.

In my Introduction to electrical engineering class (Circuits I), we cover the topic of the D'Arsonval meter.[j] For those readers who never used an analog multimeter, the basic meter consisted of a meter movement that could deflect to the full scale of its range with only microamperes of current flowing through the meter coil. So, for example, if you have an analog meter that will deflect full scale with 10 μA of current and you place that meter in series with a 10 MΩ resistor, you now have a voltmeter that can measure 10 V full scale with a 10 MΩ input resistance. Of course, the meter's windings also have a resistance that usually is part of the circuit calculation. The point of this lesson is to sensitize the student to the interplay between the circuit under observation and the tool being used to observe the circuit.

Another exercise that we go through in class is the difference between accuracy and resolution. Your digital multimeter may be able to resolve the voltage down to ±1 mV, but the accuracy of the meter may only be trustworthy down to ±15 mV on the 10 V range. As experienced engineers, we know this, but students constantly fall into the trap of accepting without question the reading on the meter.

Oscilloscopes can also be a significant perturbation to a circuit as well as a source of error in their own right due to the differences in bandwidth among the probes. I recall a student asking me why

[j] https://en.wikipedia.org/wiki/Galvanometer.

the signal amplitude of a pulse train was so low. I poked around and sure enough, the pulses should have had an amplitude around 5 V, but the scope registered below 500 mV.

I then started looking at the scope setup and I noticed two things:

1. The scope input was set for 50 Ω.
2. The probe was set at 1 × attenuation.

In effect, when the student probed the circuit node, he was hanging a 50 Ω resistor to ground on the node.

This is all part of the educational process and it is why labs are such important parts of an engineering student's education, even though they tend to complain about the time they have to spend in the labs. I often wonder if metrology should be a required course in an EE's curriculum rather than an elective course. In lecture, the students could learn the theory of measurement and their lab experiments could demonstrate the practical side of measurement instruments, such as the proper way to use them and how to understand and interpret the results.

On the digital side, the logic analyzer has always been the premier measurement tool, although that dominance may be waning (more on that in a later chapter). However, the logic analyzer (LA) can be a very intimidating instrument to learn the basics on and even more intimidating to learn how to use it to solve the really tough problems.

Trying to observe 100 or more signals going in and out of a surface-mount integrated circuit with pin spacings of 0.5 mm and clock rates of 500 MHz is not something that you can do by clipping 100 flying leads to the IC. In these situations, the measurement tool is an integral part of the system and the system must be designed from the start with the logic analyzer interface designed into the hardware. Typically, this will require that the first PC board be a "throwaway" and only used for development. Fig. 4.5 [7] illustrates the necessity of planning ahead.

This circuit adapter provides 16 input channels and 100 kΩ isolation (see equivalent load diagram in the lower right). This probing solution is recommended for normal density applications (parts on 0.1 in. centers) and where speed is not a significant issue. Keysight offers additional probing solutions and detailed application literature for probing high-speed and high-density circuits. The Keysight probes are also designed to mate to high-density surface mount connectors made by Mictor and Samtec.

However, in most cases, these solutions also require that the probing adapter be built into the PCB and be part of the planning process from the outset. Once the circuit is thoroughly debugged and characterized, the probes can be removed in the next revision of the board.

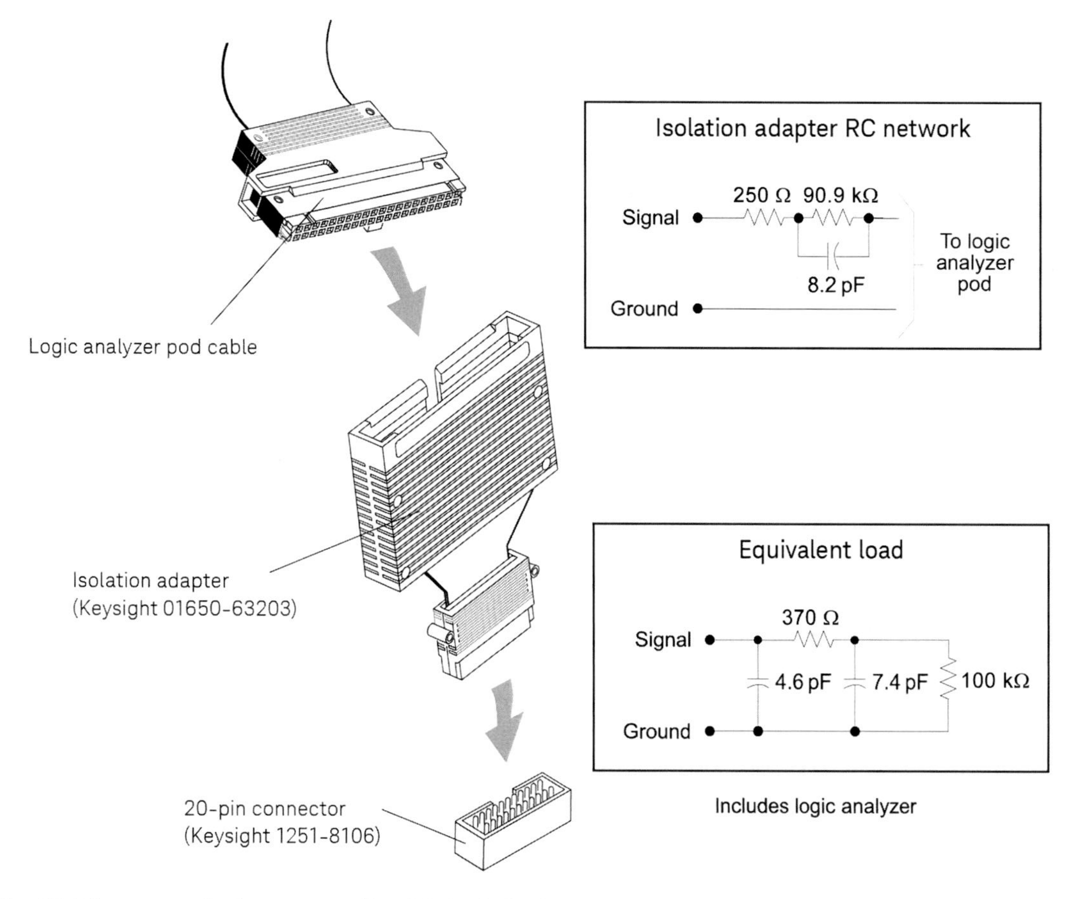

Fig. 4.5 Adaptors required to connect a Keysight Logic Analyzer to a digital system. The 20-pin connector is soldered to the circuit board and the isolation adapter provides signal isolation as shown and also mechanically interfaces the pod cable to the PCB-mounted connector. Courtesy of Keysight Technologies, Inc., with Permission.

Microprocessor design best practices

Introduction

Like software, the easiest bugs to fix are the ones that aren't there. In other words, the fewer bugs that you, the designer, designs into the system, the higher the likelihood that you can confidently blame the firmware designer for the bug (Sorry, I couldn't resist). So, in no particular order of importance or relevance are some guidelines that I teach my students in their microprocessor system design class.

If you are an experienced digital hardware designer, these rules are burned into your consciousness, or maybe they aren't. In any case, if you find these hints too elementary, just skip to the next chapter. I won't get angry, I promise.

Design for testability

Yes, I've said this before, but I can't say it enough. Design your boards so that they are testable. Bring critical signals out to easily accessible test points and liberally sprinkle the board with ground pins or pads that you can easily access. As described in the previous section, you may need to provide accessibility for other debugging tools, such as logic analyzers or JTAG ports.[k]

As you design for testability, keep in mind the perturbation that your measurement tool may have on circuit behavior, such as additional capacitive loading. One workaround is to provide a buffer gate or transistor to isolate the signal that you want to probe.

Fig. 4.6 illustrates this point. A buffer gate is being used to isolate an oscilloscope probe from the circuit. While it does address the issue of circuit loading, it adds an additional part and potential loss of synchronization (clock skew) due to the propagation delay through the buffer gate.

Consider PCB issues

The PC board does more than simply hold the components and interconnect them. The board can become part of the system in a way that is analogous to the role of measurement tools. The

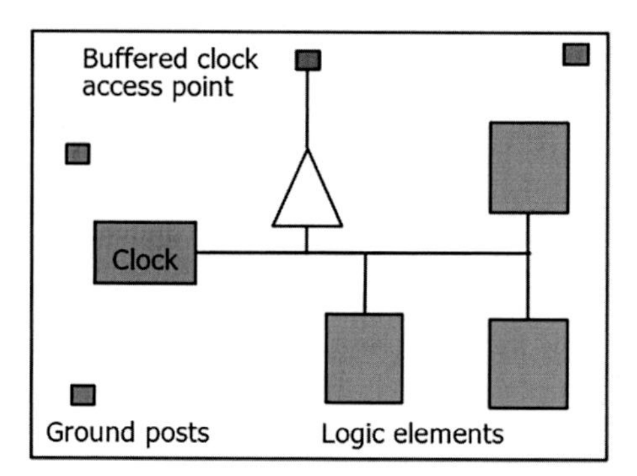

Fig. 4.6 Buffered clock test point to minimize perturbing the system under test.

[k]JTAG is an acronym for Joint Test Action Group. Originally a standard for PC board testing, it has become the de facto standard for connecting a host computer to a processor's debug core circuitry.

board can be a perturbation on the system and the part it plays generally needs to become part of your solution. Another point to consider is that your circuit may work during initial testing, but anomalies may not turn up until much later. One of a hardware designer's biggest nightmares is a circuit glitch that won't appear again for weeks, but you know it is lurking.

When cost is a major consideration, PCBs may only have one or two layers, which then puts a premium on ground and power bus management. Generally, high-speed signals want to travel over ground planes so that a constant impedance transmission line is created. However, a four-layer board with inner layer power and grounds will cost significantly more.

You will need to consider the static and dynamic electrical properties of the signal traces on the board. Obviously, the current carrying capability of the trace is important. These you can readily find online in tables from the PCB manufacturers' web sites. The thickness of the copper trace is not specified as a thickness of the copper layer (that would be too straightforward). Rather, it is expressed as a weight of copper 1 ft^2 in area. The most common copper thickness is 1 oz. copper, which translates to a thickness of 1.4 mils (0.0014 in. or ~35 μm), although 2 oz. layers are also used.

If you want to calculate the correct trace width for a given current through the trace, you can use the formula:

$$R\,(\text{ohms}) = \rho * L/A$$

where ρ is the electrical resistivity of copper, measured in units of ohm-cm; L is the trace length in cm; and A is the cross-sectional area of the trace in cm^2. At room temperature, $\rho = 1.68\ \mu\Omega\,\text{cm}$. This is noteworthy because the temperature coefficient of the electrical resistivity is positive. The resistivity goes up as the temperature goes up. This means that it is possible to end up with a positive feedback loop that turns a trace into a fuse if the current load is too great, even for an instant.

Dynamic effects add another layer of complexity. Now we need to consider the PCB material and thickness as another determining factor. Also, transmission line effects become an issue while electromagnetic interference (EMI) also comes into the equation.

I'm not an RF expert, so I'll just discuss what I've learned through having to deal with some of these issues in the past. If a fast signal is going to go to an off-board connector, such as a 50 Ω BNC or SMA connector, then it makes sense to match the impedance of the microstrip transmission line (your PCB trace) to the impedance of the cable it's driving. In order to.

The only time I ever had to deal with this issue was when I was taking it in a class, but then again, I'm not an RF designer.

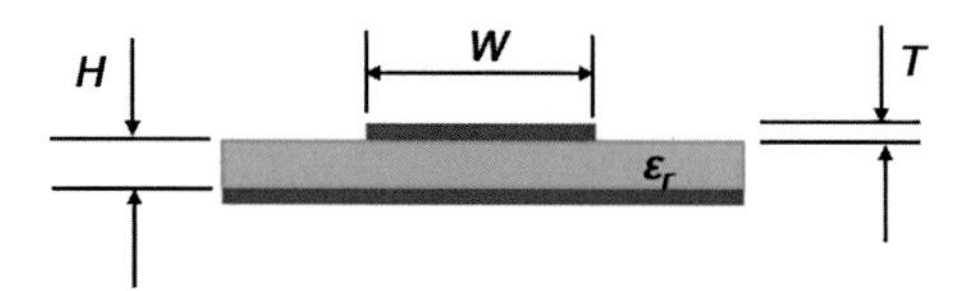

Fig. 4.7 Cross-sectional view of a microstrip transmission line (trace) on a printed circuit board.

Nevertheless, let's consider a trace on a typical PC board material (FR4) and the trace is above a ground plane, as shown in Fig. 4.7.

Here:

H = Thickness of the FR4 layer above the ground plane.

W = Width of the trace.

T = Thickness of the trace.

ε_r = Relative permeability of the FR4.

The equations of interest are as follows [8]:

$$\text{The characteristic impedance, } Z_0(\text{ohms}) = \frac{87}{\sqrt{\varepsilon_r + 1.41}} \ln\left(\frac{5.98H}{0.8W + T}\right)$$

$$\text{The distributed capacitance, } C_0\left(\frac{\text{pF}}{\text{inch}}\right) = \frac{0.67(\varepsilon_r + 1.41)}{\ln\left(\frac{5.98H}{0.8W + T}\right)}$$

$$\text{The propagation delay } T_{pd}\left(\frac{\text{ps}}{\text{inch}}\right) = C_0 \times Z_0$$

I asked a colleague on the UWB faculty who is an RF specialist[1] to choose the correct impedance for a PCB trace that will be carrying a fast digital signal. I was curious if it was arbitrary or if there was an underlying engineering principle. I'll try to synthesize his answer as follows: The goal is to avoid a situation where a pulse, such as a clock signal pulse train, interferes with subsequent pulses. This would be the case if there is a reflection at the load end of the trace and then a second reflection back at the transmission end of the trace.

The impedance of the microstrip trace should be set higher than the output impedance of the transmission gate and also at a value that would not cause the interference phenomenon just described. Thus, the transit time of the signal depends upon the length of the trace and the velocity of the signal, and the velocity

[1]Dr. Walter Charczenko, private communication.

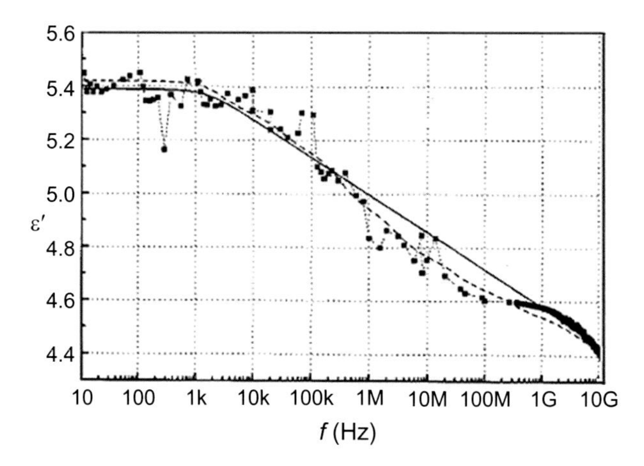

Fig. 4.8 Relative permeability versus frequency of FR4.

depends upon the impedance. So, by adjusting the impedance, we can avoid degradation of the signal integrity due to reflections.

A typical value of the relative permeability of the FR4 PCB material is 4.4, but it can vary somewhat due to variances in the manufacturer's formulation. However, we're not quite done yet because ε_r varies with frequency. Fig. 4.8 shows the variation in the real part of ε_r with frequency [9].

Referring to the figure, we can see that the frequency-dependent variation in ε_r means that the pulse will become more distorted as it travels down the trace because the speed of the Fourier components with be different and the impedance seen by each frequency component will be different.

The high-speed signals can also be attenuated along the trace because of losses due to the skin effect and in the dielectric. The higher-frequency components will be attenuated to a greater degree than the lower-frequency components. Therefore, as frequencies go higher, we need to consider other board materials such as Teflon or ceramics.

We also need to consider the reality that in a typical microprocessor-based system, we have many high-speed traces traveling in parallel over some distance. According to IPC-2251 [8]:

> *Crosstalk is the transfer of electromagnetic energy from a signal on an aggressor (source or active) line to a victim (quiet or inactive) line. The magnitude of the transferred (coupled) signal decreases with shorter adjacent line segments, wider line separations, lower line impedance, and longer pulse rise and fall times (transition durations).*
>
> *A victim line may run parallel to several other lines for short distances. If a certain combination and timing of pulses on the other*

lines occurs, it may induce a spurious signal on the victim line. Thus, there are requirements that the crosstalk between lines be kept below some level that could cause a noise margin degradation resulting in malfunction of the system.

Hand in hand with the design issues we've examined is the consideration of whether to terminate a trace on the PCB in its characteristic impedance, Z_0. Here's a good rule of thumb:

$$\text{Terminate the trace if the trace length} \geq \frac{t_R}{2xt_{PR}}$$

where

t_R = Pulse rise time or fall time, whichever is fastest,

t_{PR} = Propagation rate on the PCB, typically about 150 ps/in.

For a pulse with a rise time of 500 ps, the trace should be terminated if it is longer than about 4.25 cm.

There are several ways to terminate a trace in its characteristic impedance, and you can find resistor packs designed exactly for that purpose. According to an applications note from TT Electronics [10]:

Multi-gigabit per second data rates are now commonplace in the worlds of telecommunications, computing and data networking. As digital data rates move beyond 1-Gbit/s, digital designers wrestle with a new list of design problems such as transmission line reflections and signal distortion due to poorly selected transmission line terminators. By properly choosing a termination matching the characteristic impedance (Zo) of the transmission line, the energy in a digital transmission line signal can be turned into heat before it reflects and interferes with other forward propagating signals.

The selected type of termination is crucial to the signal integrity of high-speed digital design. In ideal designs, parasitic capacitance and inductance can kill an otherwise well-thought-out high-speed design. Care must be taken, however, when choosing a resistor for high speed transmission line termination – not just any resistor from the top desk drawer will do. A terminating resistor that matches Z_0 at low frequencies may not remain a match at high frequencies. Lead and bond wire inductance, parasitic capacitance, and skin effect can drastically change the impedance of a terminator at high frequencies. This change in impedance, and the resulting signal distortion, can cause false triggering, stair stepping, ringing, overshoot, delays, and loss of noise margin in high speed digital circuits.[m]

Traces can be terminated in three basic ways, as shown in Fig. 4.9.

[m] This article cites a primary source: Caldwell B. and Getty D., "Coping with SCSI at Gigahertz Speeds," EDN, July 6, 2000, pp. 94,96. See also reference [11], Texas Instruments Applications Note AN-903.

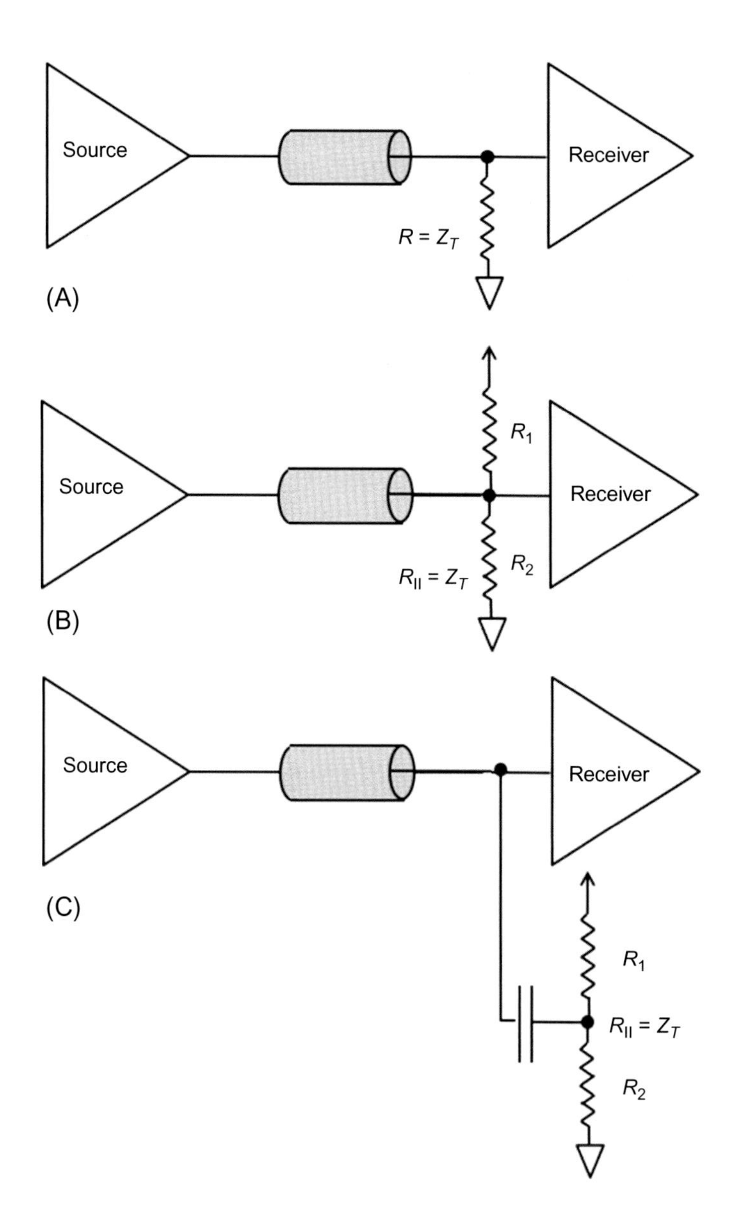

Fig. 4.9 Various forms of trace termination.

The upper drawing shows the standard termination form with a resistance to ground equal to the characteristic impedance of the trace.

The middle drawing is also a resistive termination but uses the Thévenin equivalent circuit for the termination. Although it requires two resistors per trace instead of one, it has the advantage of biasing input to the receiver at the switching point of the logic to minimize the amplitude of the current pulse during switching. The termination resistors also serve as pull-up and pull-down resistors, improving the noise margin of the system.

The bottom drawing depicts an AC termination method where the capacitor blocks the DC signal path. This considerably reduces the power demand on the signal. However, the choice of the capacitor requires some care because there is now an RC time constant to be concerned about. For example, a small capacitor value acts as an additional high-pass filter, or edge-generator, possibly adding overshoot and undershoot to the signal.

When dealing with high-speed signals, ground planes are a must. However, the decision to include ground and power planes in a design can have a major impact on the cost and performance of a design. In particular, when a microprocessor system contains analog circuitry, such as amplifiers and analog-to-digital converters (ADCs), ground management becomes an even more critical consideration.

Here's a simple example. Suppose that we are using a 13-bit ADC. This would be a 12-bit magnitude plus sign, over the input signal range of −5 to +5V. Each digital step would roughly correspond to an analog input voltage change of about 1.2mV.

On the digital side, 1.2mV is buried in the noise because we might have a noise margin of 200mV or so. But if we can see ground bounce due to digital signal switching that is greater than our analog threshold, we could be reducing the accuracy of the analog measurement. However, because digital switch tends to be fast transient noise pulses, perhaps it won't be an issue most of the time, except on the infrequent occasions when it is an issue. There are a number of good reference texts [12] devoted to proper grounding and shielding techniques for PCB designers. I'll just include one best practice that I use and that I teach my students.

Rather than having a single ground plane, split the plane into separate analog and digital planes. Many high-precision ADCs will have separate analog and digital ground pins that should be connected to their respective ground planes. This is illustrated in Fig. 4.10.

Notice how the analog section of the ADC is isolated from the rest of the digital section of the package with its own power and ground inputs. This likely carries over to the actual IC die. Separating the planes will ensure that current pulses into the ground plane do not affect the low noise requirements of the analog ground. Finally, the dot in the right diagram of Fig. 4.10 represents the connector pad of the PCB where the ground reference returns to the power supply.

One last hint with respect to ground planes. If your PC board vendor supports "thermal isolation pads," then it is a good idea

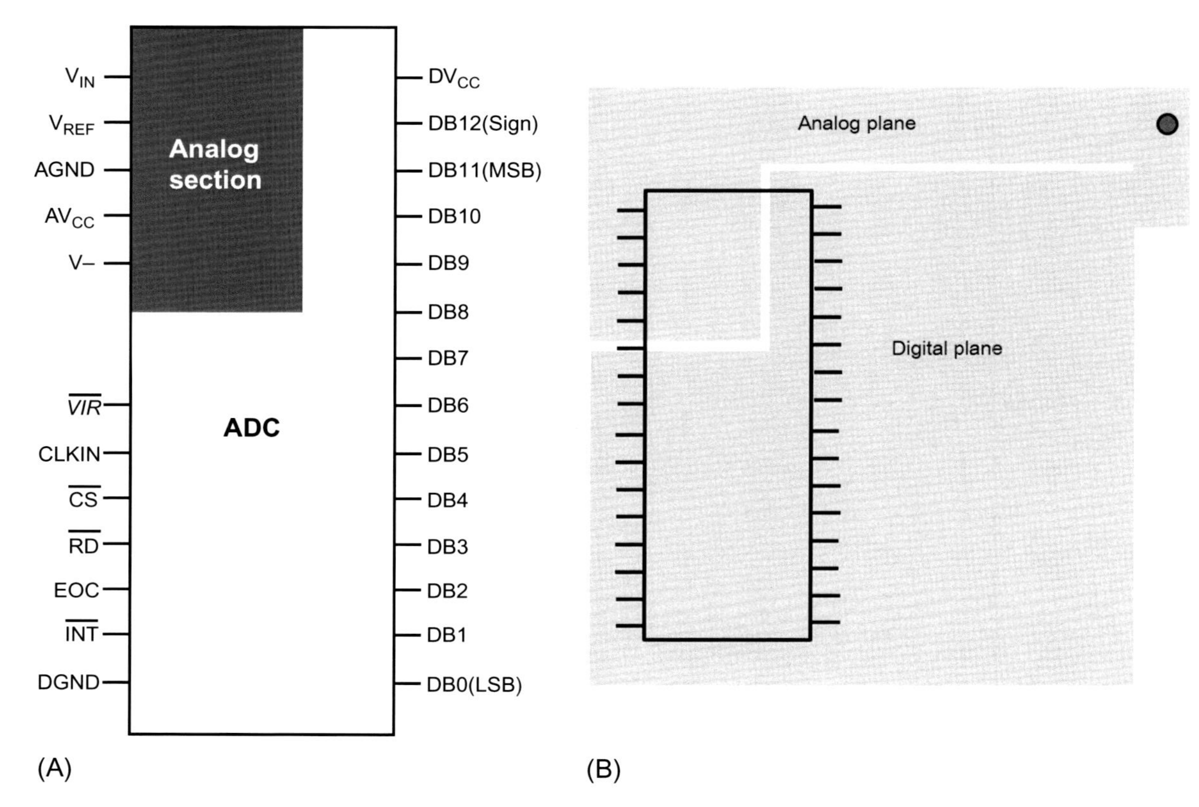

Fig. 4.10 Proper use of a split ground plane to maintain separate analog and digital grounds. Note that the ADC on the left has isolated digital and analog grounding pins as well as separate analog and digital Vcc power pins.

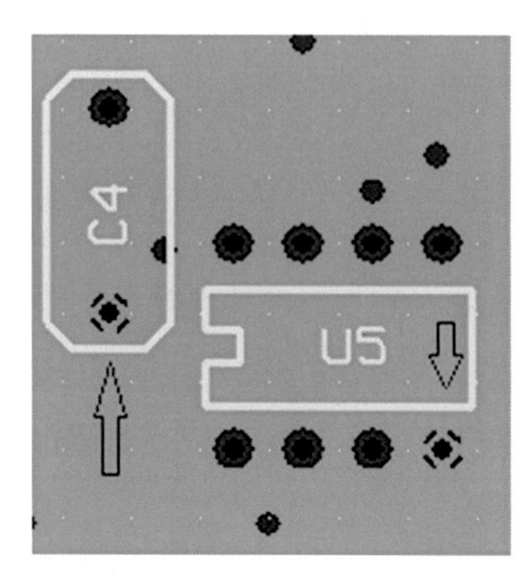

Fig. 4.11 Thermal pads on the inner layer ground plane of a PCB.

to use them. The thermal pads are indicated by the arrows in Fig. 4.11. Note how they look like a wheel with four spokes emanating from the through hole.

The purpose is to increase the thermal resistance of the pad, which, if the need arises, gives your soldering iron a fighting chance to unsolder the part from your PC board. Without the thermal pad, the copper layer is such a good thermal conductor that it will draw the heat away from the pad as fast as the soldering iron applies it.

A final note on grounding. I've included two application notes from Analog Devices[n] on proper grounding techniques in the Additional Resources section at the end of the chapter. So, if you don't want to buy a book, check out these application notes.

The last "best practice" I wish to suggest is the liberal use of power supply filter capacitors on your PCB. Students often ask me, "How many capacitors should I use?" I could launch into a long discourse, but I try to keep it simple. A good rule of thumb is one 10 μF electrolytic capacitor near the power input connector and a 0.1 μF ceramic capacitor near the Vcc input pin of each digital device.

One question I ask myself is whether a flaw in the design of a PCB is considered a bug? In other words, do I have a bug in my circuit even before I turn on the board? I would assert that is the case, so I offer this additional best practice, which is more of a guiding principle than an action. After designing many PCBs myself and helping countless students design and fabricate their

[n] www.analog.com.

own boards, I could make this observation. If a board is going to have a problem, there is a 75% probability that the problem will be mechanical, not electrical.

The experienced reader will probably laugh at this, but.... here's my list of the most common board design errors that I've seen:

- Wrong hole size
- Part interference
- Incorrect spacing for parts, particularly connectors
- Improper pin numbering on connectors, especially if they are backloaded
- Wrong pin size for connectors
- Incorrect trace width
- Failure to properly heat sink a part or provide proper cooling
- Wrong part footprint

Wrapping up

This chapter was primarily about processes rather than "here's how to find bug #7." Yes, I did include some information about a few of my favorite best practices for designing printed circuit boards because so much of what a hardware designer does centers around the PCB that holds everything together. Again, this only scratches the surface of the topic. We could spend pages with discussions of glitch detection in FPGAs[o] or similar, but that really isn't my focus.

In later chapters, a number of these specific debugging issues will be covered (I hope adequately) when I discuss how to use the tools of the trade, so I hope that you, the reader, wasn't misled by the content of this chapter versus its title.

Additional resources

1. www.debuggingrules.com: A web page devoted to debugging. Lots of good debugging war stories from engineering contributors.
2. Walt Kester, James Bryant, and Mike Byrne, ***Grounding Data Converters and Solving the Mystery of "AGND" and "DGND,"*** Tutorial Analog Devices, Inc., Tutorial MT-031, 2009, https://

[o] See, for example, Proceedings, 16th Euromicro Conference on Digital System Design: DSD 2013: 4–6 September 2013, Santander, Spain, *Glitch Detection in Hardware Implementations on FPGAs Using Delay Based Sampling Techniques.*

www.analog.com/media/en/training-seminars/tutorials/MT-031.pdf: Their list of references is worth the download by itself.

3. Hank Zumbahlen, ***Staying Well Grounded***, Analog Dialogue, Volume 46, No. 6, June 2012, https://www.analog.com/en/analog-dialogue/articles/staying-well-grounded.html. According to the author,

   Grounding is undoubtedly one of the most difficult subjects in system design. While the basic concepts are relatively simple, implementation is very involved. Unfortunately, there is no "cookbook" approach that will guarantee good results, and there are a few things that, if not done well, will probably cause headaches.

4. http://www.interfacebus.com/Design_Termination.html#b.

References

[1] A.S. Berger, A consulting engineering model for the EE capstone experience, in: Proceedings of the 2017 Annual Conference and Exposition, Columbus, OH, June 25–28, Paper 19700, 2017.

[2] A.S. Berger, Embedded Systems Design: A Step-by-Step Guide, CMP Press, 2001, p. 58. October, ISBN #1-57820-073-3.

[3] R.A. Pease, Troubleshooting Analog Circuits, Butterworth-Heinemann, Boston, MA, 1991, p. 6. ISBN: 0-7506-9499-8.

[4] A.S. Berger, Following simple rules lets embedded systems work with uP emulators, EDN 34 (8) (1989) 171.

[5] D.J. Agans, Debugging, the 9 Indispensable Rules for Finding the Most Elusive Software and Hardware Problems, AMACOM, 2002. ISBN:0-8144-7457-8.

[6] G.A. Moore, Crossing the Chasm, Marketing and Selling High-Tech Products to Mainstream Customer, revised ed., HarperCollins, New York, 1999.

[7] Keysight Technologies, Inc., Probing Solutions for Logic Analyzers, http://literature.cdn.keysight.com/litweb/pdf/5968-4632E.pdf, August, 2017.

[8] Designer's Guide for Electronic Packaging Utilizing High-Speed Techniques, IPC-2251, November, www.ipc.org, 2003.

[9] A.R. Djordjevic, R.M. Biljic, V.D. Likar-Smiljanic, T.K. Sarkar, Wideband frequency-domain characterization of FR-4 and time-domain causality, IEEE Trans. Electromagn. Compat. 43 (4) (2001) 662.

[10] TT Electronics, Digital Data Terminations A Comparison of Resistive Terminations for High Speed Digital Data, Application Note LIT-AN-HSDIGITAL, Issue 2, www.ttelectronics.com.

[11] Texas Instruments, A Comparison of Differential Termination Techniques, AN-903, SNLA034B-August 1993, Revised April, 2013.

[12] E.B. Joffe, K.-S. Lock, Grounds for Grounding: A Circuit-to-System Handbook, John Wiley and Sons, Hoboken, NJ, 2010. ISBN: 978-0471-66008-8.

5

An overview of the tools for embedded design and debug

Introduction

This is the chapter I've been looking forward to writing because this is what I sweated over for so many years. Embedded systems are unique in that they require specialized development and debug tools because the number of possible variables in a system containing untested hardware and untested software is so huge that we need these tools to be able to scope the problem into something more manageable.

The key to any of these tools is visibility. In order to fix a defect, we need to observe it in its natural habitat. There is just so much that we can do in a static and sterile environment because as much as we test and follow a disciplined development process, unanticipated events can occur. Of course, the more we deviate from a disciplined process and careen toward chaos, the greater the probability that we will have defects: defects in our code, defects in the hardware, and defects in the interplay between the two.

Sometimes we are tasked with finding and fixing defects, or just improving the performance of someone else's work. I recall a Dilbert cartoon where the engineers are sitting around a conference table at the start of a new project and (I'm paraphrasing here) Alice says, "I'd like to start the project with the traditional bad-mouthing (of) the engineer who worked on this before."

It's pretty easy to try to debug a system and blame all the problems on the engineer(s) who came before you.

After I left HP, I had a stint as R&D manager at a now-defunct embedded tool company. We had an emulation product that contained an application-specific integrated circuit (ASIC) that worked, but the documentation was so sparse and full of errors that it was virtually useless. We needed to improve the ASIC,

Debugging Embedded and Real-Time Systems. https://doi.org/10.1016/B978-0-12-817811-9.00005-3

but the original design engineer was long gone. I offered to bring him back as a consultant to help us figure out how it worked, but he was not interested, so we either had to shelve the project or start all over again. The project was shelved.

Where was I? Yes, tools. Back to tools.

The tools of the embedded designer, whether she is designing hardware, firmware, or application software, all focus on providing visibility into the behavior of the system. But visibility requires more than just seeing an event fault. It also means seeing the sequence of events that led up to the event that exhibited failure mode and also what happened after the event took place and what havoc it wreaked on the rest of the system.

How we can achieve the required level of visibility is the subject of this chapter.

Debugger

The debugger is the classic tool for observing software execution under very restrictive conditions. It is also the primary tool of the software developer, and for many software developers, it is the only tool that they'll ever need to use.

The price can vary from free, such as the GNU debugger GDB, to thousands of dollars for an industrial-strength tool chain comprising an integrated design environment (IDE), debugger, compiler, linker, loader, and other tools.

RTOS companies offer task-aware debuggers that enable a real-time operating system, RTOS, to continue multitasking and servicing interrupts while a single task may be debugged. This is a major advantage over a debugger that views the operating system as one large application and brings everything to a halt when the debugger is entered.

Because we all know what debuggers are, and likely we've all debugged code from our first CS 101 class, I won't dwell on the basics. So, let's talk about debuggers and their relationship to embedded systems.

At this point we need to define a couple of terms:

- Host computer: The computer with which the developer is directly interfacing.
- Target computer: The microprocessor or microcontroller on the system being debugged, often referred to as the target system.

If you write your code in a high-level language such as C or C++, then it would seem logical that the code could be tested, validated, and debugged strictly on the host computer just as easily, and certainly more conveniently, without the necessity of dealing with all that messy hardware stuff.

For the most part, this would be true, except when:
- Performance is an issue.
- Code size is an issue.
- The code must touch hardware.

When embedded code is being debugged on the host computer, depending upon the required level of interaction with the hardware, the developer may debug the software as if the application was being designed to run on the host. This is a pretty common occurrence and the only difference is when the high-level source code is recompiled for the instruction set architecture (ISA) of the target processor and not the ISA of the host (X86, most typically).

If there is the need for interaction with hardware, then either you need a target system to run the debugger, or you can provide a scaffold for the debugger to interact with. With a debug or test scaffold, you would write function calls instead of direct hardware calls. These function calls could return values to the calling function that would simulate what the hardware would do if it was present. Here's one example that I thought was a clean solution and relatively easy to implement.

Melkonian [1] discusses one method using macro calls in the code because macros have lower overhead than function calls. He describes the technique of replacing all direct reads and writes to hardware by corresponding macro calls as follows. In this example, he is assuming that all I/O devices are memory mapped.

```
Data read:  byte data = *(byte *)addr;
Data write:  *(byte *)addr = data;
```

These two C instructions use pointers to directly access an I/O device at the memory location in this example called "addr."

These two operations would be replaced in the code by:

```
byte data = read_hw_byte(addr); and write_hw_byte(data,
addr);
```

But the author goes one step further. Within these macros, he embeds the condition compilation of whether the code is being compiled for the host environment, in which case the macro definition becomes a function call, or for the target environment, in which case the macro defines a pointer operation. This is illustrated in this example from the author:

```
***********************************************************
*******
// Header file
#ifdef SIM
 // Simulator
 extern BYTE read_hw_byte(volatile BYTE *addr);
```

```
  extern void write_hw_byte(BYTE data, volatile BYTE *addr);
#else
  // Real target
  #define read_hw_byte(x)    (*(volatile BYTE *)x)
  #define write_hw_byte(d, x)    ((*(volatile BYTE *)x) =
  (d))
#endif
************************************************************
*******
// Simulator implementation
BYTE read_hw_byte(volatile BYTE *addr)
{
}
   intercept_read(addr);
   return *addr;
void write_hw_byte(BYTE data, volatile BYTE *addr)
{
}
   *addr = data;
   intercept_write(addr);
}
```

The author notes that there are no conditional compilation statements around the hardware access lines of code, making the code more maintainable. Using this technique, all hardware accesses have the same macro wrapper, and only the compile time preprocessor flag, SIM, denotes if the code is being compiled for the host or for the target system.

Of course, you'll still need to create the code that simulates the hardware when the call is made. Initially, you might just have it return a simple "average" value, which takes minimal time to create and enables you to keep developing your code. This is the most common approach to developing software in the absence of the hardware.

Many software developers will make use of an "evaluation board." This is usually a single-board computer, or SBC, having the processor of interest, along with a communications port, RAM, and some form of nonvolatile storage, such as FLASH memory or an SD card. These SBCs have become very mainstream of late with open-source projects such as the Arduino and TIVA from Texas Instruments as well as FPGA development boards containing hard-core embedded microprocessors, such as ARM. For example, the DE 1-SoC development board from Terasic contains an Altera (now Intel) FPGA with an embedded ARM Cortex 800 MHz dual-core processor.

Having a development or evaluation board is advantageous for a number of reasons:

- The actual embedded code can be tested and debugged, eliminating possible issues due to subtle architectural issues between the host environment and the embedded environment, Endianness being the prime example.
- Running on a real processor allows benchmarking of code performance.
- The interrupt environment doesn't have to be simulated, although some software would be needed to generate interrupts from nonexisting hardware.
- It would be possible to use real-time debug tools, such as a logic analyzer, to observe code behavior in real-time.

Still, having a development board and not the actual hardware means that hardware/software interaction must still be simulated in some way. Many processors can be set up to trap calls to nonexistent memory locations. The idea is that if your code goes into the weeds, you can limit the damage that it might be able to do. Many processors have software-initiated interrupts that automatically access specific address vectors that will point to the interrupt service routine that you would write to handle the error condition. So, if you have a divide-by-zero error or an illegal instruction code (op-code error), privilege violation, etc., the processor will vector to your error handler.

Smith [2] pointed out how this could be used as a hardware simulation method by simply using the error trap for an illegal memory access as the entry point to the simulated hardware. The advantage of this approach is that the exact low-level software that makes the hardware call does not need to be changed and the simulation code can be written in a more hardware-specific way, therefore making the simulation a potentially more accurate representation of the SW/HW interface.

If you happen to be using the ARM processor in your design and that processor shares its instruction set architecture (ISA) with the ARM processor that you might find inside an FPGA, such as the ARM Cortex inside the Intel FPGA on the DE 1-SoC evaluation board, then you have another alternative. If you have access to the Verilog or VHDL code that describes the hardware peripheral device(s) that you need to talk to, then you can build an accurate HW simulation model right there in the hardware that is your embedded processor. Of course, this only works when you are using an FPGA with an ARM core, but suppose that you are using a different processor architecture, for example something in the Intel X86 family. You can still leverage this technology, but it could be significantly more expensive.

HW/SW coverification

Suppose that you are a software or firmware engineer writing embedded code to control a board full of application-specific integrated circuits, or ASICs. You might be familiar with the chips on your PC motherboard that interface your AMD or Intel processor to the rest of the computer. These interface chips are ASICs.

Anyway, you're located in one country and the hardware team is several time zones away in another country. Think globally here. All that you have to work from is the interface specification provided to you, but of course, if you have a question you shoot an email or pick up the phone. However, they've already gone home, so you won't get your answer until the next day, and the hardware engineer might not answer your question with the greatest level of clarity, so the process goes on and time is lost.

What generally happens is that the software and hardware continue their development on parallel paths until the dreaded moment of truth comes, HW/SW integration. Then, every bug in the hardware and every bug in the software become integrated as well.

Wouldn't it be nice if the software could be continually tested against the hardware as both are being developed? In other words, coverified. That was the value proposition in the late 1990s when the technology was developed. Andrews [3] wrote a nice article looking back at the history of the technology, and it is an interesting read.

Mentor Graphics, a Division of Siemens, and Eagle Design Automation, a start-up, both developed similar products within a few months of each other. The Mentor product, Seamless,[a] is still a viable product while the Eagle product, Eaglei, died after the company went through two acquisitions. I have some skin in this game because the company that I worked for at the time, Applied Microsystems Corporation, worked with Eagle to develop a hybrid hardware/software solution and I was tasked with making the two products work together.

Before I discuss our solution, which I think was pretty cool, I will describe how the original Eaglei and Seamless products worked. Seamless has likely evolved a lot since the beginning, but the original model was a very clever innovation, so we'll look at that.

Most microprocessors have instruction set simulators (ISS). The purpose of the simulator is to take high-level code that has been compiled, assembled, and linked to create an executable

[a]https://www.mentor.com/products/fv/seamless/.

image and take it as input to a host-based program that executes the code as if it was running on the actual processor. However, it does more than just execute the code: it can also provide clock cycle-accurate instruction execution times as well as visibility into memory and internal registers. In fact, an ISS can function exactly like a debugger, except at the assembly language level.

Later improvements would interface a cycle-accurate ISS to a high-level debugger and at least address some of the issues with being able to predict software performance when a high-level language is being used. Stay with me. We're getting there, I'm in professor-mode now.

Suppose that you took the output of the ISS and fed the low-level reads and writes to memory in the simulator to another program called a "bus-functional model" that would virtually wiggle the address, data, and status pins of the processor as if the code was running on the real microprocessor. OK, got it?

Now, let's consider an ASIC under development. Assume that we're designing that ASIC using the hardware description language (HDL), Verilog. It could just as easily be another HDL called VHDL, but let's stick with Verilog for this discussion. A Verilog program looks like a C program with slightly different dialects. When you compile a C program, it compiles down to an executable code image. When you compile Verilog, it compiles down to a recipe for a silicon foundry to manufacture your ASIC.

Because manufacturing ASICs is an expensive and time-consuming endeavor,[b] Verilog environments include hardware simulators that will exercise the Verilog code and give you a cycle-accurate output of all the I/O pins on the ASIC either doing what they should do or not.

However, to test the simulation, the hardware engineer needed to create an input file of all the input pin activity on each clock cycle. These were called vectors, and each clock cycle would have an n-dimensional vector comprised of the state of all the address, data, and status lines at that particular clock cycle part of the general bus cycle.

Creating these vector tables was time consuming and the general state of affairs was that 50% of the time, a new ASIC had design errors.

Back to the bus-functional model. Here's where everything comes together, I promise. The bus-functional model took the read memory and write memory operations out of the ISS and converted those reads and writes to the input and output vectors

[b]Although FPGAs have mitigated this issue, and in fact, FPGAs were originally designed for prototyping ASICs.

as input to, and output from, the Verilog simulator. On a read operation, the bus-functional model converted the vectors to the data from the ASIC, and it would then look like real data coming back from the real hardware. That, in a nutshell, is coverification.

The advantages are pretty obvious. Once enough of the Verilog code exists for the ASIC to act like it can receive a memory write and return a memory read, even if nothing real is happening in the rest of the Verilog code, the software engineers can continuously (seamlessly?) test their code under development with the hardware under development.

Enter Applied Microsystems Corporation. Mike Buckmaster and I invented the Virtual Software Processor-Target Application Probe, or VSP/TAP [4]. The VSP/TAP allowed the Verilog simulation environment to interface with the actual hardware, or with the real processor on an evaluation board.

The VSP/TAP sits between the processor and the rest of the hardware and monitors the address, data, and status buses. When it detects a read or write to the address of an ASIC under development, it captures the bus information and activates the wait-state input on the processor, causing it to idle in place. While the processor is marking time as if it was communicating with a really, really slow memory, the VSP/TAP sends the bus operation information to the bus-functional model where it was converted to vectors, just as before with the ISS.

The VSP/TAP never really caught on, even though I thought it was a pretty neat debugging tool. Its cost and complexity as well as requiring that each microprocessor had to have a custom VSP/TAP uniquely built for it, just like other processor-specific debugging tools, made it a hard sell to a prospective customer.

What also made the VSP/TAP a problem was that most high-reliability systems had watchdog timers to monitor the health of the system and reset the processor if it detected that the software had gone astray.

If the watchdog timer has a 10-ms timeout cycle and the Verilog simulation takes 10 s to grind through to return a value, the watchdog timer has long since reset the processor many times.

It should be apparent from this (overly long) discussion that there are many excellent technologies available to provide hardware simulation so that the debugger may continue to function as a valuable tool in the embedded design process. Let's look at the pros and cons of debuggers.

We've already discussed debugging in a host-only environment, so let's consider debugging when we have a target system of some kind. It could be an evaluation board, real hardware, or something in between. In that case, the debugger is in two parts. One part of the debugger lives on the host and communicates with the user and the other lives in the target system and executes the commands that are sent to it and returns results to the host. We call the part that lives on the target system the debug kernel.

The debug kernel consists of a block of code that is either linked to the user's code when the code image is formed or loaded separately, but when loaded, it can take over the key interrupt subroutine vectors (ISR) or, finally, it can be resident in the ROM or FLASH memory of the processor.

In addition to the debug kernel, the target system needs to have a communications channel with the host. This could be any of the common serial links such as Ethernet, USB, or the old standby, RS-232.

Because the debugger now controls the interrupts, any command coming from the host can immediately stop program execution and return control to the debug kernel. The debug kernel requires that the target system hardware be reasonably stable, or at least the processor and memory systems because if the hardware is unstable, the debug kernel will be useless. Also, depending upon how the debug kernel is implemented, it can be trashed by runaway software and unlike running in a standard host environment, the debugger needs to have enough knowledge about the hardware to be able to communicate with the host.

The debug kernel implements the ***run control*** portion of the debugger. This includes:

- Examine/modify memory or registers.
- Single step.
- Set breakpoints.
- Run to breakpoint.
- Load code.

When a program is running under a debugger, single stepping, for example, it is not running real time, so the debug kernel becomes a serious perturbation to the real-time system. Also, the entire application stops running when the debugger takes over because the interrupts from the target system's hardware are usually disabled. However, a debugger can coexist with real-time target system interrupts through careful assignment of the interrupt priority levels, but this depends upon the processor having a flexible and configurable interrupt system.

However, commercially available RTOS debuggers can be task-specific. One task can be single stepping under the control of the debugger and the other tasks are happily running along.

Another general restriction on the debug kernel is that a breakpoint can't be set if the code is executing out of ROM memory.

Now, this can be mitigated if the ROM is FLASH memory, which can be written over multiple times, but if ROM is not reprogrammable by the target system, then a breakpoint can't be set.

Ever wonder how a debug kernel can set a breakpoint? When I'm lecturing, particularly when the class ends around 8 p.m. or goes over dinner, the class energy is generally at a minimum. I like to believe its physiological and not my skill as a lecturer, but who knows? To wake them up, I'll stop the lecture and ask some typical questions that they might get at a technical interview. This works every time.

Here, the question I usually ask when I'm teaching assembly language and I introduce the NOP (no operation) instruction is, "Why put in an instruction that does nothing?" After several uncomfortable moments of silence and blank stares, I can usually start a reasonable discussion.

One of the reasons is so a debugger can work. Here's why. When you tell the compiler to compile the code for debugging, it will add NOP instructions after each function. This pads the function without doing any harm, other than the size of the code image and the extra clock cycles required to pass through the NOPs. When you tell the debugger to set a breakpoint, the debug kernel springs to life and saves the existing machine language instruction at the breakpoint location in the NOP area and then replaces the instruction with a software interrupt (trap vector) that takes the processor into the debug kernel.

The state of the registers is stored, just like any other interrupt, but now the debugger can poke around and examine things. Assuming that you want to step the program by one instruction, the debug kernel replaces the original instruction and moves its trap vector into the next instruction location.

At the host end, the debugger speaks to the user and displays all the relevant information. It also maintains the databases that contain:

- Knowledge of source files.
- Knowledge of object files.
- Symbol table.
- Cross reference files.

This is the information that is needed to interface the high-level debugger to the low-level debug kernel.

Let's consider some of the pros and cons of a debug monitor (debugger). First, the advantages:

- Low cost: $0 to <$1 K.
- Same debugger can be used with remote kernel or on host.
- Provides most of the services that a software designer needs.
- Simple serial link is all that is required.

- Can be linked with user's code for ISRs and field service.
- Good choice for code development when hardware is stable.
- Can easily be integrated into a design team environment.
- Can be used with a "virtual" serial port.

The last bullet needs some further explanation. There's an old programmer's saying, "There's more than one way to skin a serial port."[c] A virtual serial port can look like a serial port to the debugger, but the implementation can be completely different than a true hardware serial port.

One implementation uses a ROM emulator to implement the virtual serial port. We'll discuss a ROM emulator in just a bit, but the Classic Comics version is a hardware device that contains RAM memory and a cable that allows it to plug into the target system where the ROM memory chip normally resides. There is a communications link from the ROM emulator to the host computer enabling ROM code to be quickly and easily loaded into the "ROM" being emulated.

The ROM emulator implements the virtual serial port. A block of memory, say 128 bytes, is allocated for the virtual serial port, basically corresponding to the ASCII character set. Any time that the processor reads from one of those locations, the ROM emulator interprets the read operation as a character transmission to the host. On the command receive side, it is often necessary to connect a "flying lead" from the ROM emulator to the interrupt pin of the processor. Not particularly elegant, but it works on target systems without communications channels.

Note that the virtual serial port implementation has pretty much gone the way of magnetic core memory, but it is a simple example of one particular way to implement a virtual serial port.

Now for the dark side of the Force:

- Depends upon a stable memory subsystem in the target, so it is not suitable for initial HW/SW integration or HW turn-on.
- Not "real time" and the system's performance may be different with a debugger present, even if the debugger is not being used.
- Difficulty in running out of ROM-based memory because can't single step or insert breakpoints in ROM.
- Requires that the target have additional services such as a communications channel, and for many cost-sensitive target systems, this is an unacceptable cost.
- If the code is not well behaved, then the debugger may not always have control of the system.

[c]Actually, I just made that up. Sorry.

ROM emulator

I hesitated when I started to discuss the ROM emulator because I wasn't sure if it was still a relevant tool, given the advancements in FLASH technology. Where we once used EPROMs and factory-programmed ROMs, we now have the ability to read and write to nonvolatile memory almost as easily as we write to RAM.

As an aside, I'm writing this text on a desktop computer that has 1TB of solid-state "disk" storage. An SSD now costs about twice the cost of a comparable mechanical hard drive of the same capacity. When SSDs started to be available, the cost premium was about 10 times the cost of the mechanical drive. Given the pace of advancement in the technology, I think we can see the cost of an SSD equal to the cost of a mechanical drive in the near future. Anyway....

I was pleased to see that ROM emulators have found some interesting niche markets, so it is fair to discuss them, even if it is only in a historical context. ROM emulation seems to be very big with the retro game enthusiasts. The old games were all ROM-based and today, to keep the games alive, enthusiasts are using ROM emulators to replace the hard-to-find game ROMs. If we ignore the copyright issues for the time being, we can see the value of plugging an emulation device into a standard ROM socket and being able to play the old games once again.

I also discovered that ROM emulation is alive and well in the test and diagnostic areas. Their focus is for test and validation, rather than new product development, but I'm sure there is still a lot of overlap. Also, many industrial control, military, and avionics applications use tried-and-true technologies that must have 25+ years of supportability, so ROM emulation would be a natural tool for systems that employ ROM-based memories to hold the operational firmware. Here's what one company, Navatek Engineering Corporation, says about ROM emulation: [5].

> *ROM Emulation is a powerful and versatile method of microprocessor testing. Introduced in 1985, ROM emulation has emerged as the technique of choice for microprocessor test and diagnostic applications. The board is tested by replacing the boot ROMs on the Unit Under Test (UUT) with memory emulation pods. Each pod handles 8 bits of the data bus, so processors from 8 to 32 bits can be controlled with 1 to 4 pods (even the most advanced CPUs such as the Intel Pentium generally use only an 8-bit boot path). The emulator takes control of the UUT by resetting the processor and, under the test program's control (monitor program),*

it can then exercise all functions on the board. Synchronization with the UUT is automatic and requires no additional hardware or connections.

ROM emulators have several key advantages over other tools that are specific to one microprocessor. This comes about because ROMs tend to have very standard I/O pin descriptions and these descriptions tend to stay the same even as the ROM capacity increases and more I/O pins are added in order to handle the additional address lines.

Fig. 5.1 illustrates this point. As the ROM capacity increases, the relative position of the I/O pins stays more or less in the same position. It isn't until the EPROM with 16 data lines, 271024, that the pinout changes significantly. From the point of view of the ROM emulator manufacturer, this makes life very easy, as only a

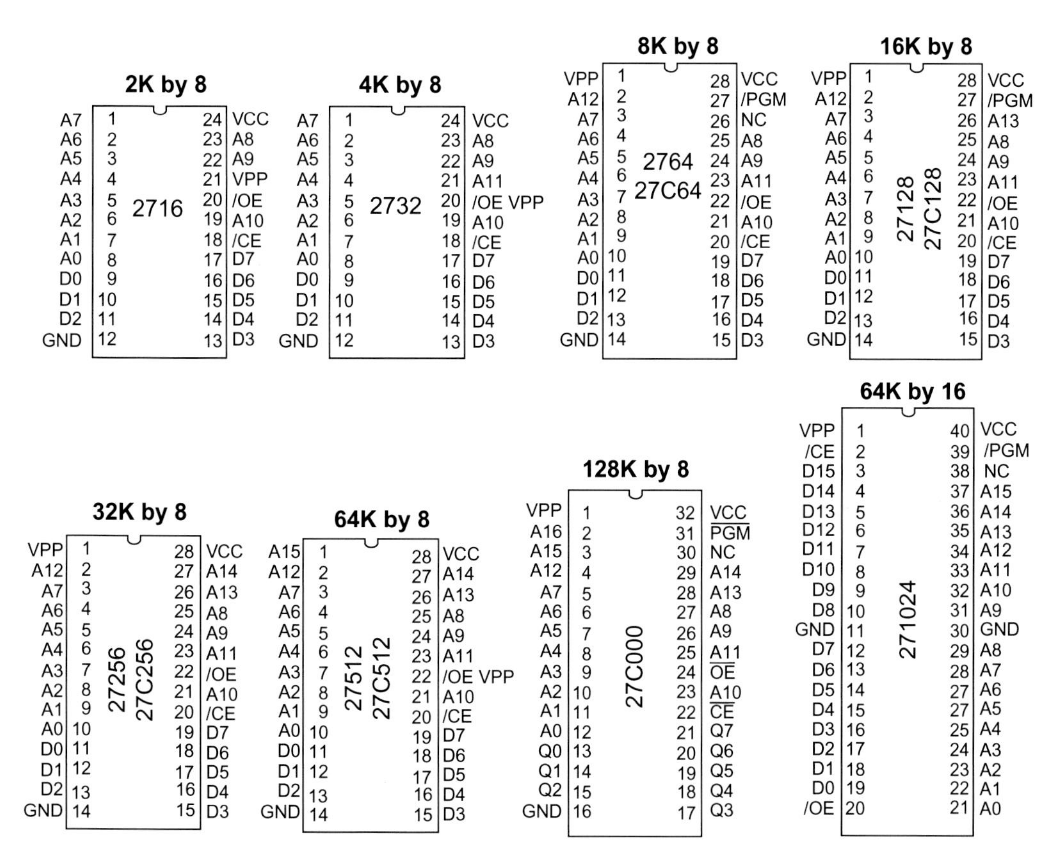

Fig. 5.1 Standard pinout configurations of various capacity EPROMS. Preprogrammed parts would have the same configuration with the exception of the programming pins such as VPP and /PGM that would not be available.

few different cabling configurations need to be developed to support a wide range of devices from a wide range of manufacturers.

The ROM emulator also provides several other possible benefits besides a simple way to load instruction code into an embedded system. These can be enumerated as follows:

- Virtual serial port for target systems that lack a debugger communications port.
- Can implement debugger breakpoints in ROM.
- For applications where a product must be certified to meet "mission critical" confidence levels, a ROM emulator with appropriate trace memory can record memory addresses of all instructions that were accessed during testing, thus providing evidence that all instructions were executed and all memory addresses have been executed (no dead code regions).
- With additional trace memory, program execution can be traced just as if a logic analyzer is in the ROM emulator.
- With sufficient memory in the ROM emulator, entire memory regions can be instantaneously swapped out, enabling a simple way to turn debugging on and off, or to have system test code be coresident with instruction code.

In other words, the ROM emulator is an extremely flexible tool for designing and testing embedded systems. However, the Achilles heel of a ROM emulator is of course:

- The program code needs to be stored in ROM, either to execute out of ROM or to be moved to RAM at boot-up time.
- The ROM should be socketed, although this might not be an issue in a lab prototype.
- The ROM should be a standard type.
- The ROM needs to be placed in an accessible location on the PCB.

Given these constraints, a microcontroller with internal memory would not be a good candidate for a ROM emulator unless the internal memory can be bypassed during development and external memory used instead.

Architecturally, the basic ROM emulator is rather straightforward. The RAM is dual-ported so that either the local control processor or the target system can access it. The RAM should have a sufficiently low access time so that the delays introduced by any multiplexer or buffer circuitry will not be an issue.

Additional logic may be necessary, depending upon the number of I/O possibilities that must be accounted for. A local control processor manages the system and communicates with the host computer. The local processor should be able to read standard object file formats such as S-Records or Intel Hex. The processing power of the local processor will depend upon

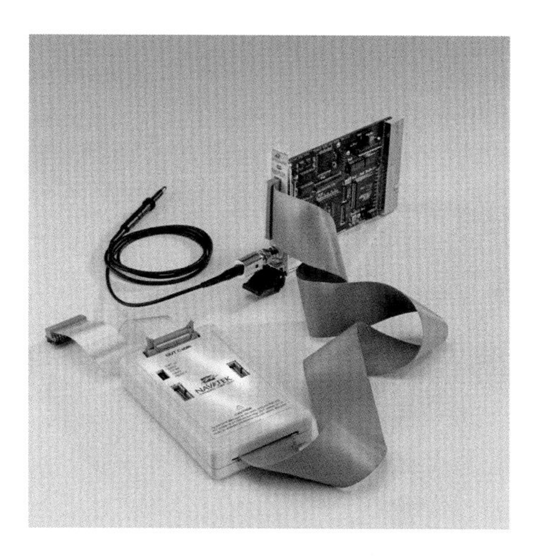

Fig. 5.2 ROM emulation test system. Courtesy of Navatek Engineering Corporation.

the feature set of the ROM emulator, but otherwise, it's a straightforward design problem.

Fig. 5.2 is the Navatek NT5000 ROM Emulation Test System. This particular ROM emulator is designed to plug into an industry-standard PXI test system chassis, such as those manufactured by National Instruments[d] or other vendors. The plug-in card also includes a logic probe for nodal diagnostics such as cyclic redundancy character (CRC) checks or signature analysis.

Other ROM emulators use USB or Ethernet to connect to a host computer. Unlike ROM emulators used for software development, the NT5000 is used for diagnostic testing of a microprocessor-based board.

Aside from the retro game enthusiasts and the test and verification manufacturers, there are still a few small companies selling ROM emulators. One of note is the PROMJet from EmuTec Fig. 5.3.[e] Of note is that the PROMJet supports a number of ROM footprints besides the DIP standard and other standard 27XXX family footprints. According to their website [6]:

> *Direct connect cables are available for DIP, PLCC, TSOP, PSOP and BGA sockets. Custom cables can be created for other socket*

[d]www.ni.com.

[e]http://www.emutec.com/.

Fig. 5.3 PROMjet ROM emulator. Courtesy of EmuTec.

> *configurations… The PROMJet SPI/LPC option allows it to support the increasingly popular SPI (1-bit Serial Peripheral Interface) or LPC/FWH (4Bit Low Pin Count/Firmware Hub) FLASH memories. A wiring adapter is only needed to adapt the SPI FLASH 8-pin SOIC footprint to PROMJet 50-pin header. For LPC/FWH FLASH devices, PROMJet supports the 40-pin TSOP as well as the 32-pin PLCC footprints.*

You might be wondering why you would need a ROM emulator for FLASH memory because FLASH memory can be reprogrammed in-circuit. According to the EmuTec website: [7].[a]

> *It is a development tool for embedded systems that eliminates the need for programming Flash memory during firmware development cycle. It replaces the Flash memory of the system under development allowing the user to load, examine, modify, view and patch the program code directly into PROMJet's emulation memory.*

The PROMjet system includes more than 50 different FLASH memory adapters for connecting to different FLASH memory (or ROM) footprints. It also has the ability to output interrupt signals to the target system and trigger signals for an oscilloscope or logic analyzer.

I don't have any financial arrangements with either EmuTec or NavaTek Engineering. I just think ROM emulators are very versatile and useful tools for embedded development and debugging when the system includes nonvolatile memory that is external to the microprocessor or microcontroller.

Logic analyzer

While I was planning and researching this book, I had a number of informal discussions with former colleagues about the future of logic analysis. Some thought that the viability of the logic analyzer was on a downward glideslope, citing the higher level of integration of systems on silicon (SoC), the quality of simulation tools, and the greater prevalence of high-speed serial buses.

Counterarguments were that as long as digital systems need to be designed, hardware tools such as oscilloscopes and logic analyzers will be necessary. I think I'm in the latter camp, but I think everyone is correct from their particular perspective. Digital systems continue to evolve and the tools that are required to support their design, test, and validation processes will also continue to evolve alongside them.

Therefore, let's take a leap of faith and assume that the logic analyzer will still be around and viable, at least for the useful life of this book, and see how it works and what can be done with it.

The logic analyzer is capable of making two basic types of measurements depending upon the relationship of a timing clock to the desired measurement.

Timing measurement: The timing measurement is the easier of the two to understand because it is conceptually closest to the digital oscilloscope. The trace is captured in much the same way with the clock being internal to the logic analyzer. When the scope or LA is triggered, the internal clock determines the frequency with which the signal is captured and stored in memory. Just like scopes have signal bandwidths from 10 MHz on up to the multi-gigahertz (depending on the depth of your pockets), logic analyzers follow the same frequency limits.

However, oscilloscopes rarely have more than four input channels while logic analyzers can easily have several hundred simultaneous inputs. On the other hand, the logic analyzer is a 1-bit digitizer while the scope may easily have 12 or more bits of resolution. So, the logic analyzer and the oscilloscope trade-off signal fidelity, or resolution in one case, for many input channels in the other.

If the clock speed is sufficiently faster than the clock in the system under test, then the logic analyzer would be able to display and measure the timing relationships between various signals, enabling set up and hold measurement, for example. As a simple example, suppose that your logic analyzer is capable of 2.5 GHz capture rates, such as the Keysight 16861A 34-Channel Portable Logic Analyzer. The period of a 2.5 GHz clock signal is 400 ps. So, if you are trying to determine the time difference, Δt, between the positive clock edge and the chip enable (~CE) input on your memory going low, the 16861A will provide Δt to a resolution of ±400 ps.

State measurement: This is where the oscilloscope and the logic analyzer part company. In the state measurement, the clock that determines when data is captured is in the target system, rather than the LA. Therefore, the "state" of the system is captured on each clock edge of the system clock.

The state measurement displays the data as a table, much like a spreadsheet, with the number of the clock cycle, or relative time, defining each row in the table and each column defining one of the signals being measured.

To illustrate this, I'll use two screenshots from my microprocessor design class (B EE 425). To introduce the logic analyzer, I designed a simple microprocessor board using the venerable Z80 processor. Before you roll your eyes, let me explain the rationale. The Z80 has one really useful feature that makes teaching logic analysis really effective. Everything that you see on the buses is exactly what is being processed. There are no caches, no prefetch queues to confuse the students when branches occur. It is delightfully inefficient, but a great teaching tool.

Fig. 5.4 is a timing diagram for the Z80 that you could find on thousand of websites. We see the relationship between the clock and address, data and status buses for a memory read cycle and a memory write cycle.

For these measurements, the clock would be internal to the LA. This is what a timing measurement would look like on your screen. Now consider Fig. 5.5.

This is a state chart for an 8051 microcontroller and it shows the *state* of each of the address, data, and status signals. This time the clock would be one of the inputs to the LA and data would be captured only on the positive or negative edge of the clock, depending upon how the LA was set up. The gray row represents the trigger position and the red (dark gray in print version) and blue (light gray in print version) colored rows are cursors that can be set in the display.

Going down the first column on the left, you see the elapsed time between each clock pulse. In this instance, the logic analyzer uses data compression to increase the number of states it can hold in its memory, so sometimes it seems that the clock period is 120 ns and other times it is 60 ns. This happens to be a trace display for the LogicPort logic analyzer that we use in our microprocessor teaching lab.

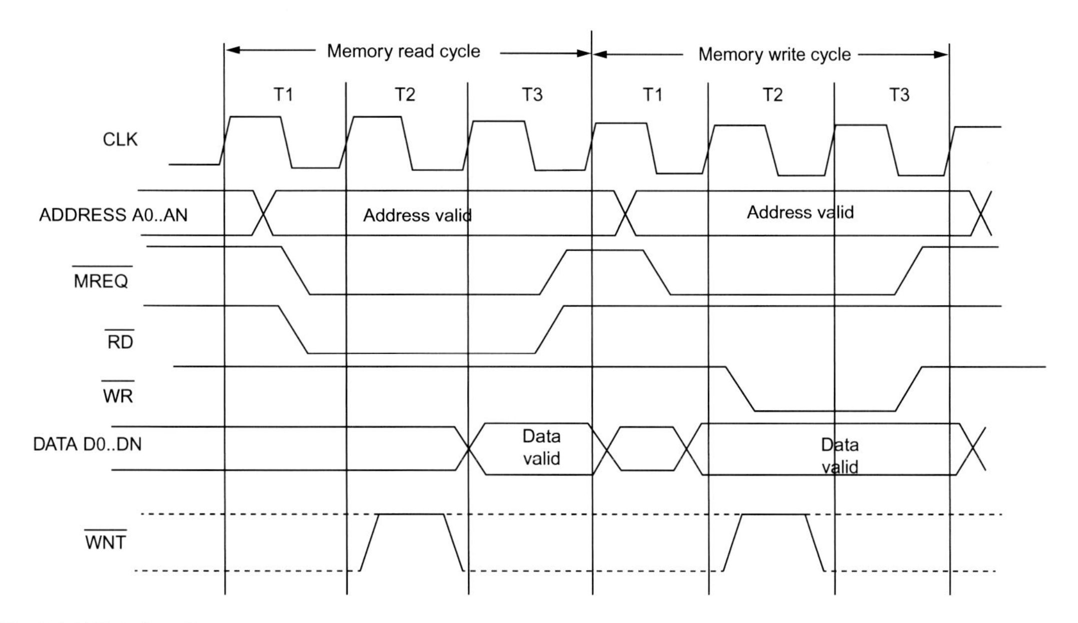

Fig. 5.4 Z80 timing diagram.

Relative to Reference	XTAL1	RD	WR	PSEN	ALE	AD[7..0]	A[15..8]	Latched A[7..0]	RST	INT0	INT1
T+0ns	1	1	0	1	0	00h	02h	D1h	0	1	1
+60ns	0	1	0	1	0	00h	02h	D1h	0	1	1
+180ns	1	1	0	1	0	00h	02h	D1h	0	1	1
+260ns	0	1	0	1	0	00h	02h	D1h	0	1	1
+380ns	1	1	0	1	0	00h	02h	D1h	0	1	1
+460ns	0	1	0	1	0	00h	02h	D1h	0	1	1
+580ns	1	1	0	1	0	00h	02h	D1h	0	1	1
+660ns	0	1	0	1	0	00h	02h	D1h	0	1	1
+780ns	1	1	0	1	0	00h	02h	D1h	0	1	1
+860ns	0	1	0	1	0	00h	02h	D1h	0	1	1
+980ns	1	1	0	1	0	00h	02h	D1h	0	1	1
+1,060ns	0	1	0	1	0	00h	02h	D1h	0	1	1
+1,180ns	1	1	0	1	0	00h	02h	D1h	0	1	1
+1,210ns	1	1	1	1	0	00h	02h	D1h	0	1	1
+1,260ns	0	1	1	1	0	00h	02h	D1h	0	1	1
+1,380ns	1	1	1	1	0	00h	02h	D1h	0	1	1
+1,400ns	1	1	1	1	1	CDh	02h	D1h	0	1	1
+1,410ns	1	1	1	1	1	CDh	02h	C1h	0	1	1
+1,420ns	1	1	1	1	1	CDh	02h	CDh	0	1	1
+1,460ns	0	1	1	1	1	CDh	02h	CDh	0	1	1
+1,580ns	1	1	1	1	1	CDh	02h	CDh	0	1	1
+1,660ns	0	1	1	1	1	CDh	02h	CDh	0	1	1
+1,780ns	1	1	1	1	1	CDh	02h	CDh	0	1	1
+1,790ns	1	1	1	1	0	CDh	02h	CDh	0	1	1
+1,860ns	0	1	1	1	0	CDh	02h	CDh	0	1	1
+1,980ns	1	1	1	1	0	CDh	02h	CDh	0	1	1
+2,000ns	1	1	1	0	0	CDh	02h	CDh	0	1	1
+2,010ns	1	1	1	0	0	C0h	02h	CDh	0	1	1
+2,020ns	1	1	1	0	0	D2h	02h	CDh	0	1	1
+2,060ns	0	1	1	0	0	D2h	02h	CDh	0	1	1
+2,180ns	1	1	1	0	0	D2h	02h	CDh	0	1	1
+2,260ns	0	1	1	0	0	D2h	02h	CDh	0	1	1
+2,380ns	1	1	1	0	0	D2h	02h	CDh	0	1	1

Fig. 5.5 State diagram for an 8051 processor. Courtesy of Intronix.

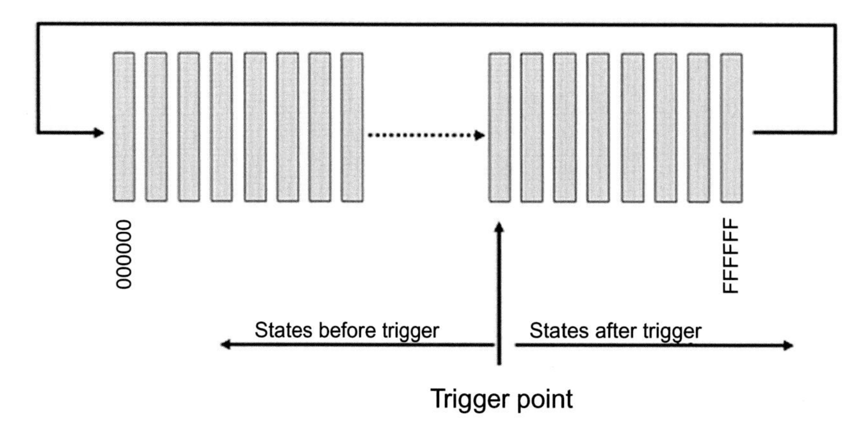

Fig. 5.6 Architecture of a logic analyzer circular trace buffer. The trigger point defines the memory address when the LA stops overwriting the previous data in the buffer.

The second and perhaps more significant way that a logic analyzer and an oscilloscope differ is the triggering capability. Both the logic analyzer and the digital oscilloscope use the concept of a circular trace buffer, or trace memory. In the days of the analog oscilloscope, the trigger signal caused the electron beam to travel across the screen, displaying the time-varying voltage versus time. If the oscilloscope was capable of waveform capture, or waveform *storage*, it was accomplished by analog techniques that we don't need to discuss here but got me my job at Hewlett-Packard's Colorado Springs Division designing the electron optics for the storage cathode ray tube in the HP1727A oscilloscope.

Digital oscilloscopes changed all that. With the advent of very high-speed analog-to-digital converters, a different architecture became possible. With digital storage, the oscilloscope or logic analyzer is constantly running. The trace memory system is a circular buffer, so that as the instrument runs, it continually fills the circular buffer and overwrites the previous information. The trigger now serves a different function. The trigger signals when to stop overwriting the previous data, not to start taking data.

The implications of this are quite profound because with a circular buffer, you are able to see what happened before the trigger signal. In other words, in negative time.[f] It also allowed you to see what led up to the event that caused the trigger to occur. This is shown schematically in Fig. 5.6.

[f]I vividly remember reading a science fiction short story about an engineer who invents a transistor that turns on several picoseconds before a control voltage reaches the gate. It was called an anticipatory transistor.

Assume that the trace buffer in Fig. 5.6 is 256 bits wide and 16M (2^{24}) states deep. Assume that the trigger point is set for the middle of the trace buffer. Wherever the buffer pointer happens to be in the buffer when the trigger condition is met, the system will stop overwriting data at the memory address=ADDR (Trigger point)—0x800000. Let's assume that the trigger point happens to occur exactly when the buffer pointer is at memory address 0x800000. This means that when the trace storage system reaches address 0x000000, it will stop overwriting the data already in the buffer. In most logic analyzers that I'm familiar with, the trigger point is continuously variable from the beginning of the trigger (all states after the trigger occur until the buffer is filled) to the end of the buffer (every state that occurred up to the trigger point).

It is apparent with an oscilloscope that brings a great deal of flexibility to the instrument, particularly in the case of mixed signal oscilloscopes that combine the conventional oscilloscope inputs with 16 digital channels, like a logic analyzer. Current Keysight oscilloscopes and logic analyzers can be linked to perform coordinated measurements.

However, even with a deep memory, setting a simple trigger condition for the LA will often fill the buffer memory with 16 million useless states and not capture the true event of interest. A good example of this is a software loop. If you set the trigger at the address of the entry point into the loop, the actual event of interest within the loop might not take place until many hundreds of thousands of cycles in the loop, quickly filling the trace buffer with useless information. This is why logic analyzers have very sophisticated and complex trigger specification capabilities to help winnow down the exact sequence of events that led up to the trigger condition of interest.

Fig. 5.7 shows the trigger point set-up menu for the LogicPort logic analyzer that we use in our microprocessor teaching lab. By current LA standards, this is fairly rudimentary, but it illustrates the general concept. I'll give you an example of a more complex system in a little while. Note that there are two possible definitions of a trigger condition, denoted as Level A and Level B. Levels A and B can be used independently or sequenced, or logically combined as follows:

- When Level A is satisfied.
- When Level B is satisfied.
- When either Level A or Level B is satisfied.
- When Level A and Level B are satisfied in any order.
- When Level A is satisfied and then Level B is satisfied.

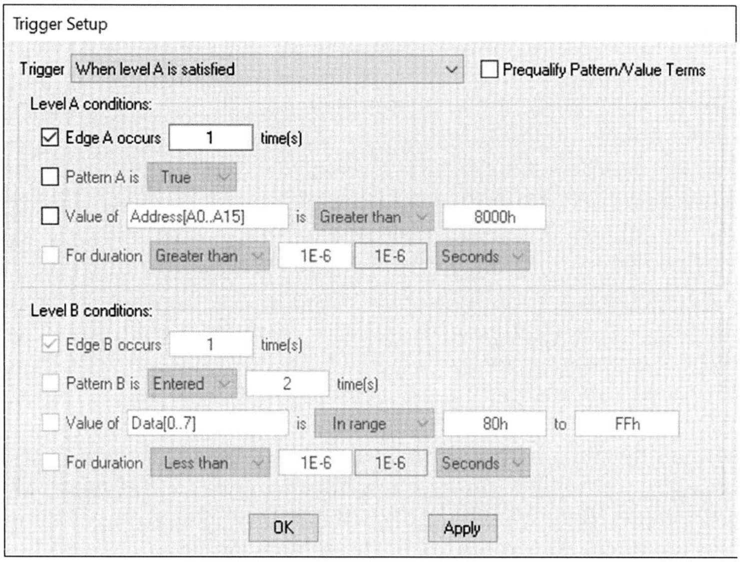

Fig. 5.7 Screenshot of the trigger set-up window for the LogicPort logic analyzer.

In this example, the Level A condition is satisfied as soon as an edge (Edge A) is detected. Here, Edge A is defined in the timing window. In this example, the analyzer trigger is set to trigger when the ~RESET input to the processor goes high and the processor comes out of RESET. However, if you look at the possible combinations of address, data, and status signals that could be used to set a pattern, or a counter to detect a time interval or a count, some of the usefulness of the trigger condition should become apparent to you.

Consider the last two bullet points. What we have here is a sequence of events. We could look at this as a simple state machine, or that Level A is used to qualify the occurrence of Level B. The more feature-rich the LA, the more capability is built into the triggering system to isolate the one event that is deeply buried within the behavior of the software and will only become apparent if a unique sequence of events occurs.

Fig. 5.8 illustrates a far more complex trigger sequence.

Here, host-based software that supports the logic analyzer is able to consume the compiler output files and linker information to associate variable and function names with address and data values. The advantage is obvious. The software engineer can enter the variable names with the same abstraction level as the code that is being developed. Notice that the diagram shows two sequence levels, indicated as Level 1 and Level 2. The logic

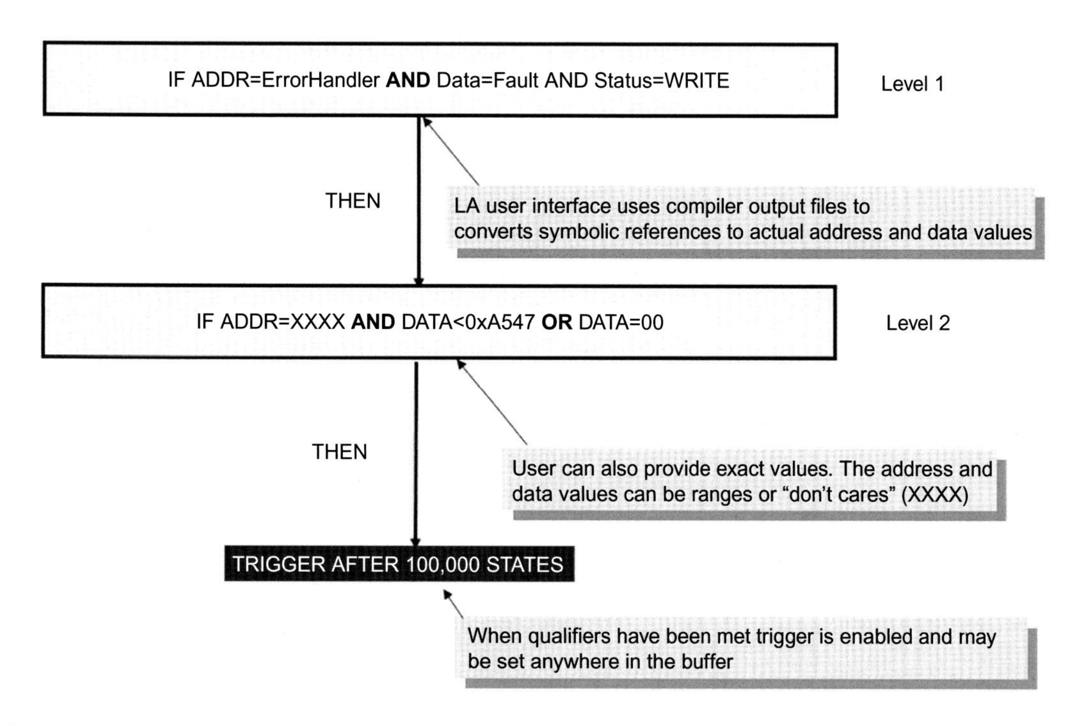

Fig. 5.8 Complex trigger sequence for a logic analyzer including symbolic names for the addresses and variables of interest extracted from the compiler output files and linker maps.

analyzers that I am familiar with[g] have up to eight sequence levels that could be used to follow a sequence of software events to isolate a fault.

In Fig. 5.8, we see the logical conditions that must be true in order to advance to the next state. Finally, after Level 1 and then Level 2 are satisfied, the LA will wait another 100,000 clock cycles before triggering. What isn't shown in this diagram is that at any level, the logical state condition can also cause the state sequence hardware to return to a previous state and try again. This should seem very natural to you if you studied finite state machines in an introductory digital design class. If we stop to think about the electronics, it is pretty impressive that this sequencing hardware is ripping along at the speed of the system clock.

Another capability available in the more feature-rich logic analyzers that is worth mentioning is the ability to only record a small number of states every time the trigger condition occurs. Rather than filling the trace buffer when the trigger sequence occurs, the

[g]We have 12 HP 16600A logic analyzers in our microprocessor teaching lab. While no longer manufactured, they still work extremely well and due to their flexibility, they are used to teach advanced microprocessor design classes.

trigger qualifier can be told to only record a few states and then start over again, waiting for the next occurrence of the trigger sequence.

This capability is extremely useful in situations where a global variable is being clobbered, or there is fault causing the stack to overflow. By setting the LA to only look for writes to the variable, the trace can show exactly which functions or RTOS tasks are accessing the variable. For a stack overflow, the LA can be set to record only stack push operations followed by stack pop operations at the exit point of functions that don't line up.

Just like being able to enter a complex trigger specification in terms of the high-level symbolic addresses and symbolic data, the trace can also be postprocessed so that the real-time software execution flow can be displayed in terms of the high-level source code.

Fig. 5.9 illustrates how postprocessing can be used to raise the abstraction level of the LA output in order to show the code execution flow in either assembly language, upper trace, or C instructions, lower trace.

```
Label:  Address  Data              Opcode or Status                 time count
Base:     hex    hex                   mnemonic                      absolute
after   004FFA   2700    2700  supr data rd word               ------------
+001    004FFC   0000    0000  supr data rd word               + 520       nS
+002    004FFE   2000    2000  supr data rd word               +    1.0    uS
+003    002000   2479  MOVEA.L 0001000,A2                      +    1.5    uS
+004    002002   0000    0000  supr prog                       +    2.0    uS
+005    002004   1000    1000  supr prog                       +    2.5    uS
+006    002006   2679  MOVEA.L 0001004,A3                      +    3.0    uS
+007    001000   0000    0000  supr data rd word               +    3.5    uS
+008    001002   3000    3000  supr data rd word               +    4.00   uS
+009    002008   0000    0000  supr prog                       +    4.52   uS
+010    00200A   1004    1004  supr prog                       +    5.00   uS
+011    00200C   14BC  MOVE.B  #000,[A2]                       +    5.52   uS
```
(A)

```
Label:       Address         Data        Opcode or Status w/ Source Lines        time
Base:        symbols         hex              mnemonic w/symbols                  rel
-009    sysstack:+003FC2     0738     0738  supr data wr word                     520
-010    sysstack:+003FC0     0006     0006  supr data wr word                     480
+011    main:main+00000A     01AA     01AA  supr prog                             520
       ##########main.c - line   104 ##############################################
           initialize_system();
-012    main:main+00000C     4EB9   JSR     |_initialize_sys                      480
+013    main:main+00000E     0000     0000  supr prog                             520
+014    main:main+000010     114A     114A  supr prog                             480
       ##########initSystem.c - line     1 thru    38 ############################
       void refresh_menu_window();
       void
       initialize_system()
       {
+015     |_initialize_sys     4E56   LINK    A6,#00000
```
(B)

Fig. 5.9 Two postprocessed logic analyzer traces. The upper trace, (A) illustrates simple postprocessing to highlight the assembly language instructions and the lower trace, (B) shows a fully postprocessed trace showing function names.

Students in my microprocessor class who are interested in an independent study project can upload the LogicPort LA trace buffer to an Excel spreadsheet and then write a Java program to disassemble the trace back to the assembly language mnemonic level. Unfortunately, postprocessing beyond that is a bit over their heads.

Lastly, the ability of the logic analyzer to only save one or more states every time the trigger specification is met makes the logic analyzer a very useful software profiling tool. This can be quite straightforward to implement and in fact, it was one of my lab experiments when I taught embedded systems to computer science students. In this lab, the students used the HP16600A logic analyzers to record certain data patterns that were introduced into their C code as immediate data writes to a specific memory address at the entry and exit points of functions.

The logic analyzer was set up to look for the pattern 0x55555555 at the entry of a function and 0xAAAAAAAA at the exit of the function. Each time the state was stored by the LA, it was automatically time-stamped by the LA.

Fig. 5.10 shows a screenshot of the trace output. In this lab experiment, the students were tasked with comparing the execution time for a processor with caches on and caches off. The actual rationale was to demonstrate how the structure of an algorithm could impact the performance of a processor. We had a contest for the student who could construct an algorithm that demonstrated the greatest difference between cache on and cache off performance. The best stress-test algorithm achieved better than a 10 × performance difference.

If this processor was running under an RTOS, the LA could be used to gather statistical data on minimum, maximum, and average execution times for functions with multiple tasks running concurrently.

Let's wrap up this section on the logic analyzer with a summary of the pros and cons of using a logic analyzer as a tool for debugging embedded systems.

Logic analyzer benefits

- Probably the most versatile of all the tools a microprocessor system designer would need.
- High-performance units can be quite costly, but inexpensive units work quite well.
- Operates in timing or state mode giving both the hardware designer and the software designer a real-time picture of what the system under test is doing.
- Complex triggering allows the processor to hone in on very complex event sequences.

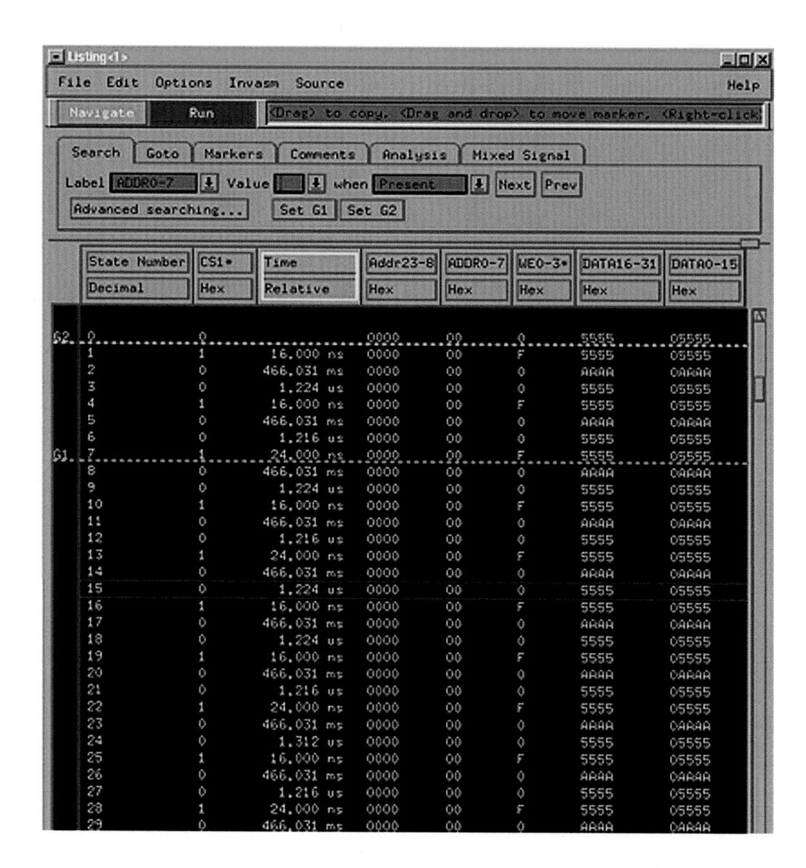

Fig. 5.10 Trace of a performance measurement using the HP16600A logic analyzer.

- Postprocessing the trace data allows the software engineer to understand the system behavior at the appropriate level of abstraction.

Logic analyzer issues

- Loses relevance when the observed behavior at the I/O pins of the processor does not reflect the internal operation of the microprocessor:
  - Example: Instruction cache turned on.
- Limit of usability is when both possible destinations of a BRANCH instruction are entirely contained within the cache.[h]

[h]This is not exactly true, as I learned when I was the R&D Director for X86 products at my last employer before I joined UW, Applied Microsystems Corporation (now defunct). They had an excellent technology that enabled a real-time trace to be reconstructed with very minimal I/O outside the processor caches. I never learned exactly how it was done, but it was very impressive nonetheless.

- Operation depends upon connecting many tiny probes to the signal lines of the system:
  - Modern processors have many tiny I/O pins spaced as close as 0.005″.
  - Must use fragile and expensive adapters to connect to the processor under test.
  - For best results, PC board should be fabricated with LA interfacing in mind.

In summary, the logic analyzer has unique capabilities to observe microprocessor systems running in real time provided that the issues of information visibility can be addressed. The LA was king when we had separate microprocessors and memory systems, like a modern PC. As technology advanced and embedded systems evolved to systems on a chip, SoC, it is a fair question to ask whether the logic analyzers will still be relevant going forward. My belief is that there will be a long glide slope simply because we will still need tools to debug and repair embedded systems that are still out in the field and replacement systems being designed using the classical embedded architecture.

What will be necessary is to anticipate the need for such tools and design in the capability of connecting them to a system. This could be as simple as adding a high-density connector to a board, or even easier, design the capability of soldering such a connector to the board in case the board needs to be debugged.

In-circuit emulator

For many years, the in-circuit emulator (ICE) was the premier development and debug tool for embedded microprocessor systems. That preeminence has been waning for a long time and there are few remaining companies that still provide tools that we would recognize today as an in-circuit emulator.

I thought about this for quite a while because I was deeply involved in ICE design and development for many years. I still have several HP 64700A family units that I built from scrap parts. I once saw one of the units I had helped to design in a surplus electronics store in Silicon Valley. I think it was on sale for $25 and the original list price was around $10K. I felt like buying it just to rescue it from such an ignoble fate.

Fig. 5.11 shows the author (me) in my younger days in a publicity photo for an advertising brochure for the HP 64700A family of ICE units. That's my cubicle in the R&D lab of the Logic Systems Division of then HP (now Keysight) in Colorado Springs, CO.

Fig. 5.11 The author using an HP 64700 in-circuit emulator to debug an oscilloscope control card.

An HP Vectra PC with dual 5.25″ floppy disk drives sits on my desk and the ICE unit is on the shelf, with a cable plugged into a socket where an MC68000 processor would normally reside.

The PC communicates with the ICE over RS-232C. This was actually quite nice because all the commands were text-based, so the ICE could be used without the GUI in a simple terminal window.

An ICE is an aggregate, integration, and expansion of three debug tools: a logic analyzer, a debugger, and a ROM emulator. Of course, there's more to it than that, for example, ROM emulation is only a small part of what the internal memory system could do. In theory, an ICE unit could be designed for almost any microprocessor or microcontroller. Larry Ritter and I wrote an article [8] a long while back in which we basically argued that what an emulator integrates could be separated into a loosely coupled, rather than tightly integrated, system. The loosely coupled system was just the logic analyzer, a debugger, and a ROM emulator. This is still a valid argument, particularly with on-chip debugging, as the microprocessor can be emulated using the in-circuit microprocessor rather than needing a separate one, as would be the case with a traditional emulator.

This issue is really how much transparency must you have in order for the ICE to meet your needs? Transparency generally refers to how closely to the actual system performance will the system behave with an ICE unit attached. Transparency has always been the Holy Grail of emulation designers. Here's a simple example.

A customer using the HP64753A emulator for the Z80 processor complained that the system worked with the Z80 installed, but it was unstable with the emulator installed. After some back and forth, I found out that their PCB did not use a ground plane, but rather had ground and power traces on the board.

The Z80 emulator used some pretty hefty buffers to drive the cable to the board and I suspected that they were seeing ground bounce due to the poor grounding on the board. I suggested that they build a socket extension adaptor with 330 Ω resistors between the emulator plug and their socket on the board. This would limit the current pulse (I hoped). It worked and another customer problem was solved.

The demise of the ICE was started by the advancement in technology that we take for granted today. I'll be discussing on-chip debug circuitry in a later chapter, but that was the culprit. Advances in IC technology, in particular the evolution of hardware description languages such as VHDL and Verilog, meant that the chip designer didn't have to sweat over every logic gate in the design. Today, transistors are free. Almost every microprocessor today comes with an internal debug core. The extra overhead of these unused gates in a production chip is rather irrelevant. Once the debug core took hold, chip manufacturers saw that internal debug functionality could be a selling point, just like chip performance.

One of the selling points of the ICE was that you had total control of the processor, even if the memory system was not stable. In fact, the ICE could be thought of as a really expensive single-board computer because the processor, memory, and debugger were already in place and a communications link to the host computer was already available. With an on-chip debug core, this is no longer an issue because the core will communicate with the host without operating memory. Even more compelling, so many microcontrollers have on-chip memory that external memory is not needed.

So, why am I droning on about an obsolete technology? I started to ask myself that, but I realized that by explaining what the classic ICE was capable of doing and how it does it what it does, I would better inform the reader about how their debugging solution would measure up against one of these units.

The ideal situation to use an ICE would be either a processor that is socketed, capable of being socketed in some way, or at least having some way to turn off the processor that is onboard. Several older processors that were soldered to the board had a disable input pin that could cause it to go dormant with high-impedance (Hi-Z) outputs. In this state, another processor could essentially be piggybacked over the dormant one via a cabling arrangement of some kind, or even a parallel connector that would take a standard, high-density cable.

HP emulators required that a mounting fixture be glued to the chip package so that the cable connector could press down on the I/O pins. This was quite impressive when you look at an IC with 0.5 mm pin spacing and 100 or more I/O pins.[i]

Assuming that the processor could be either disabled or removed, the design of the board had to be such that the socket was accessible and properly oriented so the cable could reach it without interference. Because many of the embedded systems consisted of multiple PC cards in a card cage, the only way to access the control card was if it could be put on an extender and moved out of the chassis. Sometimes that was possible, sometimes not.

OK, we have a processor that we can somehow substitute for and it is located so we can access it. Now what? Ideally, the memory system is external to the processor, or at least some memory is external. The degree to which the memory is external determines the degree to which the logic analyzer portion of the emulator will be useful. Even with on-chip caches, postprocessing software was surprisingly good at figuring out the actual code execution flow, even with on-chip caches and a knowledge of what the source code and object code looked like.

Even better, if some overhead could be tolerated, then tags could be placed in the code at strategic locations to generate noncached writes to external memory so that they could be trapped by the logic analyzer. A tag could be thought of as a low-intrusion printf(). It generally consisted of a single assembly language write instruction to an external memory address that could be monitored by the LA. Sometimes this is possible, sometimes not. Of course, if you could take the performance hit (that pesky transparency issue once again), you could disable the caches and run without them. Let's say you can. Now you have a system that you can emulate.

The key to making an emulator work is the notion of the context switch. We understand that in terms of an operating system, a generalized emulator works basically the same way. The linkage to the logic analyzer functionality and the emulation memory comes through the use of the nonmaskable interrupt, or NMI. The NMI becomes shared between the target system and the emulator. Either can generate it. The NMI causes the processor to stop execution and take an interrupt vector to the NMI interrupt service

[i]These cables were so fragile and expensive that one of our customers, who shall go nameless, kept the spare cables in a locked closet and only the manager had a key. If an engineer broke one, they had to see the manager to get a replacement.

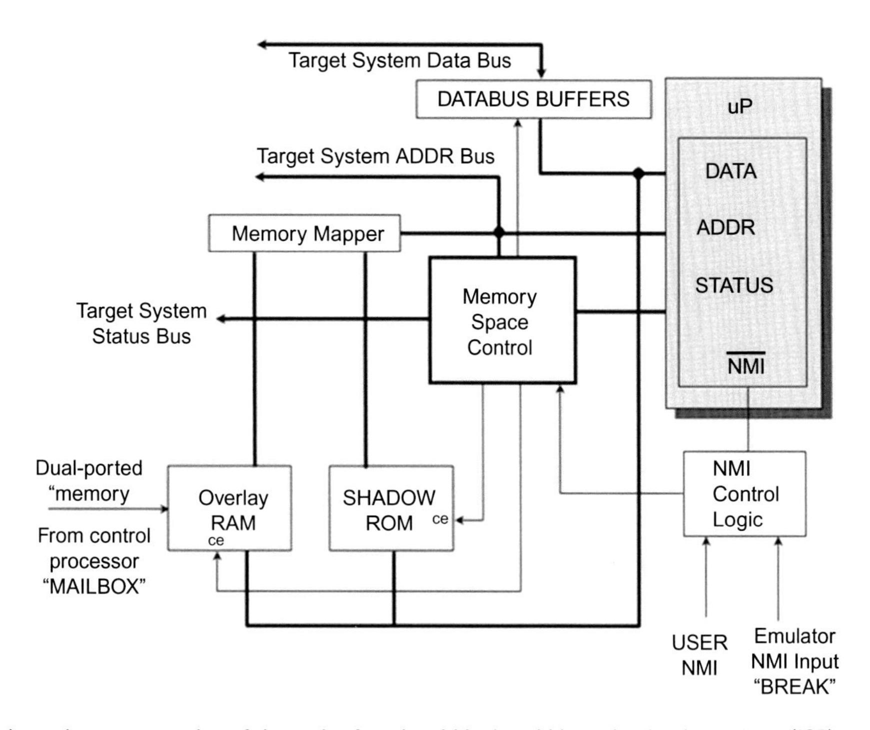

Fig. 5.12 Schematic representation of the major functional blocks within an in-circuit emulator (ICE).

routine (ISR). This is where the context switch occurs. Fig. 5.12 is a very simplified overview of the basic context switch mechanism.

The left-pointing arrows represent the I/O pins from the processor going to the target system. This then "become" the microprocessor in the target. There are two blocks of memory in the emulator. The shadow ROM block contains the debugger and is switched in when the processor sees an emulator-generated NMI. The overlay memory can be substituted for the target system memory on a block-by-block basis, typically as fine grain as 1 Kb. A spare block of memory can be assigned in the emulator to emulation memory and the rest can be assigned to the target system.

Let's say that you assign 1 Mb for your application code that should be running out of ROM on the target system. You assign this address space to the overlay memory and give it the attribute of ROM. By assigning it to be a ROM, any attempt by the program to write to this address space will generate an automatic break into the background debugger. Overlay memory is typically very fast RAM, and generally an expensive part of the emulator. It has to be fast because of the switching that needs to take place when a new address is analyzed to see if it should go to the target system or to the emulation memory. Other attributes for the overlay memory block might be "protected," read/write, write only, read only, and ROM. The overlay memory can also have extra bits assigned to each byte so that software validation can be implemented.

One of the most important functions of validation is to make sure that all the software is exercised. An extra bit can be set every time that memory location is accessed. After running the program, the percentage of bits set to 1 can be measured.

Let's trace the operation of the run control circuit. An NMI is detected by the processor. Assuming the interrupt was generated by the emulation hardware, the ISR will send the processor into its debugger, which is located in emulation memory, not processor memory. Thus, the debugger does not use any of the processor's address space. This is called a "background" debugger. Another advantage is that the target system does not require a communications link to the host PC because the emulator takes over that function. Once in the emulator's debugger, the operation is identical to a software-only debugger. You can single step or set breakpoints, peek and poke memory, registers, etc. However, another plus is that you can set breakpoints in ROM; a software-only debugger can't do that. The reason is that emulation memory is very flexible and overlay RAM can be assigned to look like ROM, but it can be modified by the emulator to implement breakpoints.

NMI execution breaks can also be generated by the logic analyzer or by the user. In each case, the emulator takes over control of the processor. Thus, the integration of the debugger/run control, real-time trace, and overlay memory (ROM emulation) centers around the breakpoint capability of the system and how flexible it is.

Emulators can still be relevant when the processor has a debug core. With a debug core, all the run control and some level of trace can be implemented, depending upon the feature set of the debug core. For example, the Nexus 5001 Forum has defined a scalable debug core standard, starting with a run control-only system for simple processors up to additional real-time trace capability.

With a debug core, the architecture of the emulator changes. It is no longer necessary to have a second emulation processor that takes over for the target system processor because the visibility into the processor behavior is defined by the debug core. All that is required is a connector on the target system that links the host computer to the debug core and some interface device that has USB or Ethernet on one side and the debug protocol on the other side. These devices are typically much less than $100, making this configuration an attractive alternative to the $10 K emulator. Also, the debug cores tend to be consistent, at least within a company's product line. This means that one inexpensive interface device is all that is required to interface to any of the microprocessors or microcontrollers in the company's portfolio. On the other hand, emulators were very specific to each microprocessor or microcontroller. An emulator for an MC68331 would not, in general, work for the MC68332. This meant that every microprocessor required a different emulator, which was wonderful for the emulator vendors, but not for the customers.

Perhaps the biggest drawback was the dependence on third-party vendors to supply the emulators in a timely manner for each microprocessor that required them. This led to a love-hate relationship between the semiconductor manufacturer and the emulator manufacturer.[j] The semiconductor manufacturer wanted emulators available when the first silicon chips were available for their customers to evaluate. The emulator manufacturer wanted to wait and see if a chip would be successful in the market before it would be willing to invest the resources to develop an emulator for it.

Another point of conflict was how does one define success. For the semiconductor manufacturer, this is sales volume. One customer buying a billion parts was ideal. For the emulator manufacturer, sales volume was irrelevant; their key figure of merit was the number of design wins for that chip. Each design win represented a potential customer. This was rather irrelevant for the first-tier semiconductor vendors such as Intel and Motorola, who dominated the embedded marketplace. Any microprocessor or microcontroller that they released would be popular enough that emulation support was guaranteed. Other manufacturers were not quite so well supported. Because I worked for AMD, and my job was to coordinate the third-party tool vendors who supported our chips (primarily the Am29000 family and Am186 family, a derivative of the Intel 186 family), I was constantly visiting customers and tool manufacturers coordinating the release of AMD's microprocessor and microcontrollers with the availability of support tools from our third-party vendors.

In Chapter 8, we'll look at the technology and evolution of on-chip debugging support. For now, this is a good place to end the chapter.

[j]I speak from personal experience because I worked for both an emulator manufacturer (HP) and for a semiconductor manufacturer (AMD).

Additional resources

1. https://www.youtube.com/watch?v=Q3Rm95Mk03c.
   Good explanation of how debuggers work.

References

[1] M. Melkonian, Software-Only Hardware Simulation, vol. 164, Circuit Cellar, Inc., 2004, pp. 58–67. No. 3, March.
[2] M. Smith, Developing a Virtual Hardware Device, vol. 64, Circuit Cellar INK, November, 1995, pp. 36–45.
[3] J. Andrews, HW/SW co-verification basics: parts 1–4, in: HW/SW Co-Verification Basics, May 23, 2011. https://www.embedded.com/design/debug-and-optimization/4216254/1/HW-SW-co-verification-basics–Part-1—Determining-what—how-to-verify, parts 2–4 follows.
[4] M.R. Buckmaster, A.S. Berger, System and Method for Testing an Embedded Microprocessor System Containing Physical and/or Simulated Hardware, US Patent #6,298,320, October 2, 2001.
[5] http://navatek.com/wordpress/nt5000-rom-emulation/.
[6] http://www.emutec.com/.
[7] http://www.emutec.com/flash_memory_usb_emulator_hardware_promjet.php.
[8] A. Berger, L. Ritter, Distributed emulation: a design team strategy for high-performance tools and MPUs, Electron. News (1995) 30. May 22.

6

The hardware/software integration phase

Introduction

The hardware/software (HW/SW) integration phase is the point in the project when untested software and untested hardware are brought together for the first time. Power is applied and either the hardware catches fire, nothing happens, or, miracle of miracles, there are signs of life. Of course, there are gradations within this nightmare scenario, but the key characteristic, the one major issue faced by the developers of embedded systems, is that you have more unknown variables than most other categories of product designs.

Consider the problem without my doomsday scenario. A software developer, writing applications for a PC or smart phone, has a standard platform with known APIs. In this scenario, bugs are overwhelmingly due to errors in the application code being written at the time. Now consider the same scenario when the hardware platform is not standard and has not been thoroughly tested and wrung out. This is the issue we face when we are developing new embedded systems and mitigating this issue is the bread and butter that keeps embedded tool vendors alive.

The HW/SW integration diagram

The classic model of the embedded system lifecycle is shown in Fig. 6.1. I'm certain that if we extracted this graph from every tool marketeer's slide deck, we could create one massive coffee table book. Fundamentally, they all tell the same story:

Phase 1: Product specification: This is where the idea for a product comes from. It might be from the marketing department, from

Debugging Embedded and Real-Time Systems. https://doi.org/10.1016/B978-0-12-817811-9.00006-5

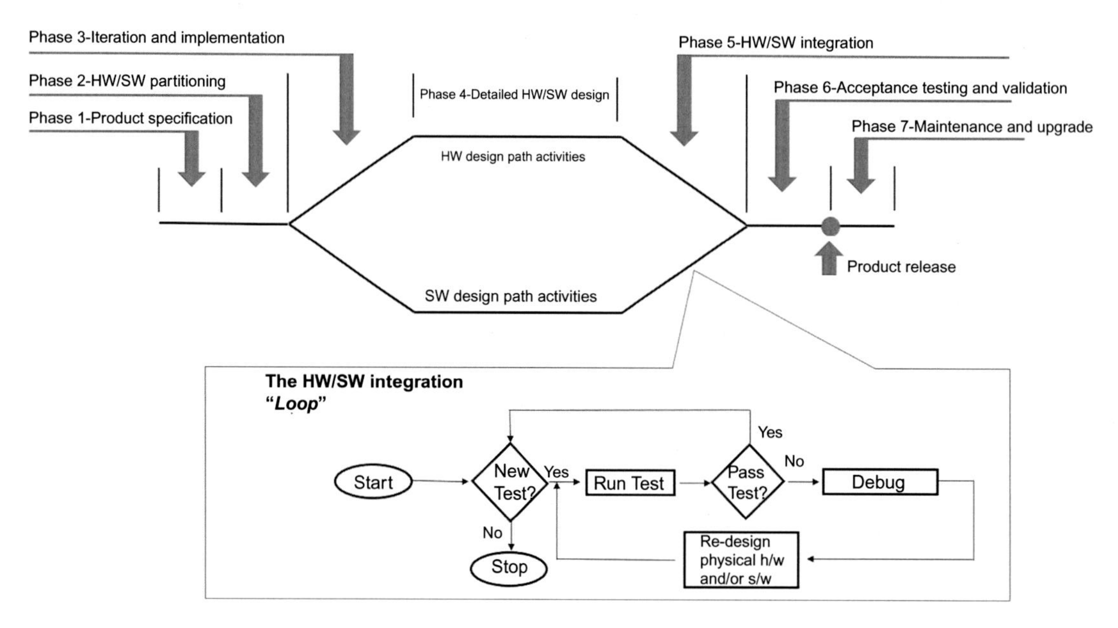

Fig. 6.1 Classic embedded systems life cycle model.

sales, from R&D,[a] from a customer, or from a competitor. Wherever the germ of the idea comes from, this is the phase where it is fleshed out and a set of specifications is derived. Many companies will also create a list of "musts" and "wants." The "musts" list represents the features that are needed for it to be successful and the "wants" are features that would be added if time and/or resource pressures permit.

This is also the phase where the concepts of "internal specification" and "external specification" are developed. The external specification is what the customer sees. These are the specifications that you may need to validate with customers through various forms of market research, such as a focus group or through customer visits.[b]

[a]At HP we called this the "next bench syndrome." If you could convince the engineer at the next bench that you have a good idea, perhaps there is a possible product there somewhere.

[b]Strictly speaking, you wouldn't use a customer visit to check on a feature set, but if you had previously met with that customer and discussed the problems that they faced and needed solutions, then bringing your product idea back to them for validation and comment might be a good idea. See McQuarrie [1].

The internal specification is the roadmap for how the product will be designed. It includes details about the processor, the memory, clock speed, hardware and software tools, customer use environment (office, home, outdoors, military, hospital, industrial, etc.), manufacturing cost goals, development schedule, and anything else that the design team needs to begin work on the project.

Like most of the other phases to follow, Phase 1 is iterative. There is, or should be, a fluidity that allows the various project stakeholders to provide input. Stakeholders are any group that may be impacted by the product as they must execute on it. For example, I once was working on a project and during one of the regularly scheduled reviews, both manufacturing and QA let it be known that a particular semiconductor vendor was no longer viable due to manufacturing shortages and higher than normal part failures in other products. Because I wasn't planning to use the vendors' parts, it wasn't an issue, but it could have been. Most of the time, any well-respected vendor's parts are acceptable and usually aren't called into question except when the part is only available from that manufacturer. A sole-sourced part generally got special attention because, as should be obvious, if that part goes away, the product must be redesigned or obsoleted, and products in the field might not be repairable.

Many products designed for industrial or military applications have useful lifetimes of 25 years or more. Products that are intended to be used in these applications need to have special provisions for extended life support. One situation that I was aware of when I worked for a semiconductor manufacturer was when an aerospace company decided to use our product in an avionics application. In order to get the contract to supply them with processors, we had to do a number of production runs ahead of any sales of the part, and then put those processors away in a safe place should we stop production of the part anytime in the future.

What about debugging Phase 1? Does it make any sense at all? Can you have bugs in the specification of something? I would advocate for a more general definition of debugging embedded systems to include any defect that arises in any phase of the product development life cycle. That definition should include a process misstep, an erroneous assumption, bad marketing data, and judgment errors because we are human and product definition is not an exact science.

Let's consider some possible defects in the definition phase of an embedded microprocessor system. I'll just cover a few that I am aware of or was personally involved in.

The case of the nonstandard hard disk drive interface

Hewlett-Packard introduced its Vectra Portable CS computer in 1987, and it contained a floppy disk drive and a 20 MB, 3.5″ hard drive. The company had high hopes for the portable, but it failed in the market. Here's the description from the HP Computer Museum [2]:

> *The Vectra Portable CS was the portable version of the Vectra CS. The Portable CS had a large LCD screen as well as CGA adaptor for use with an external monitor. The Portable was offered in two mass storage configurations: dual 3.5 inch (1.44 MB) floppy disc drives - P/N D1001A, or a floppy disc drive and a 20 MB hard disc drive - P/N D1009A. The Portable CS did not succeed due to its large size (much larger than the Portable Plus), relatively high price and non-standard media (3.5 inch discs).*
>
> *The Portable Vectra CS was introduced on September 1, 1987. It was discontinued on May 1, 1989.*

The company had high hopes of selling the hard drive as an OEM product to other computer manufacturers because it was one of the first hard drives out in the 3.5″ form factor. Rather than adopt the industry standard *integrated drive electronics* (IDE) interface, developed by Compaq and Western Digital in 1986, they designed an interface that used a 40-pin connector, but otherwise, was entirely different. As a result, the drive was never adopted by other manufacturers and HP discontinued manufacturing soon after. The lesson? Standards matter.

The last gasp of the vector display

The Colorado Springs Division of HP produced oscilloscopes and OEM displays for other HP divisions. These displays were vector displays. These displays were called vector displays because they drew images on the screen by writing a line, or vector, from one point to another. Text was created in the same way.

We were convinced that vector displays, despite the fact that they were expensive, would never be replaced by raster displays such as the display in a TV or computer monitor because of the phenomenon called "the jaggies"; raster displays had jaggies and displays did not. The jaggy was the slight stair-step effect that was apparent in a line on the screen that was not exactly horizontal or exactly vertical. Even with smaller and less-expensive raster displays becoming more prevalent, our display group continued to push products with vector displays, even as the market for these displays dried up. The lesson? Don't continue to go with old technology when all indicators point to its demise.

Underpowered emulator card cage

I can recall another poor decision that resulted in a failed product. It involves the HP 64000 product family. The original HP 64000 was a standalone workstation. Because most of our customers were migrating to UNIX workstations and HP had just purchased a workstation manufacturer (Apollo), we felt it necessary to develop a product that would interface with a UNIX workstation. The product was a card cage box with a control card and power supply that would take the existing HP 64000 personality cards. A decision was made to use an 8-bit MC6809 microprocessor to control the box rather than the much more powerful MC68000, 16/32-bit processor. The reason for the decision was that having the MC68000 processor in the box might allow a user to run UNIX on the card cage itself. Because of this decision, the box was so underpowered that the performance was terrible with long delays in data transfer and response.

Feature creep and the big customer

This is the classic marketing story about the sales or marketing engineer who returns from a visit to a major account and insists that a new feature must be added to the current product under development because that client wants it. This usually results in "feature creep" because the R&D team was uncertain about the product that they were actually trying to invent, so they decided to cover their rear ends (CYA) by adding a lot of features that really weren't necessary and just added cost and time to the development schedule.

For 5 years while I was a project manager at HP, I served on the Project Management Council, a group sponsored by HP's Corporate Engineering. We met several times a year and tried to come up with initiatives that could disseminate the best practices within the many worldwide divisions of HP.[c] One of the Project Management Council members[d] undertook a study of several successful and failed projects. The study highlighted 10 areas that were deemed crucial to the product's success.

[c] This was before the Agilent spin-off and the breakup of HP into HP Enterprises and HP, Inc.

[d] Edith Wilson was an HP Manager who originally undertook this study while at HP and later used the results as a basis for her MS degree at Stanford University [3].

Figs. 6.2 and 6.3 show the results of the study. These images were sanitized and were taken from my lectures for a course, The Business of Technology, that I taught for a number of years.

The data speaks for itself. The key factor here is the data from the first row of the matrix, "User's needs understanding."

While all the successful projects showed that they had attempted to have a good understanding of what their customers needed, only two of the six project teams for the failed projects attempted to do the same thing.

We can't say for certain that these deficiencies directly led to the marketplace failure of projects A through F in the same way that we can point to a circuit design flaw or a coding error and assign a causal relationship to the defect. The reason is that we are trying to assess blame as a defect, and that is questionable at best. However, continuing the development process with a design flaw that results from questionable decisions in the specification phase is still a flaw. Fortunately, finding the defect is easy. Nobody buys the product. Unfortunately, the defect can only be fixed by replacing the defective product with a better one.

Phase 2: HW/SW Partitioning: Partitioning is the process of deciding what is going to be done in hardware and what will be done in software. This is not always an easy decision. In order

+ = Project team performed task **- = Project team did not perform task**	Project A	Project B	Project C	Project D	Project E	Project F
User needs' understanding	-	+	-	-	-	+
Strategic alignment and charter consistency	+	-	-	+	-	+
Competitive analysis	-	-	+	+	-	+
Product positioning	-	-	+	-	+	+
Technical risk assessment	+	-	+	-	+	-
Priority decision criteria list	+	+	+	-	-	-
Regulatory compliance	-	-	+	+	+	+
Product channel issues	-	-	+	+	+	+
Project endorsement by upper management	+	+	-	+	+	+
Total organization support	+	+	-	+	+	+

Fig. 6.2 Chart of six projects that failed in the market due to defects in the extended design's initial project activities [3].

+ = Project team performed task – = Project team did not perform task	Project G	Project H	Project J	Project K	Project L	Project M
User needs' understanding	+	+	+	+	+	+
Strategic alignment and charter consistency	+	–	+	+	+	+
Competitive analysis	+	+	+	+	+	+
Product positioning	+	+	+	+	+	+
Technical risk assessment	+	+	+	+	+	+
Priority decision criteria list	+	+	+	+	+	+
Regulatory compliance	+	+	+	+	+	+
Product channel issues	+	+	+	+	+	+
Project endorsement by upper management	+	+	+	+	+	+
Total organization support	+	+	+	+	+	+

Fig. 6.3 Chart of six projects that were successful in the market. Note how the extended design team carried out almost all the activities not done by the project teams in Fig. 6.2 [3].

to introduce the concept to my students, I often use this example. "How many of you are gamers?" I ask. I'll get 1 or 2 students in a class of 25 raise their hand. I then ask them how much they paid for their graphics card in their game machine. Typically, it's in the $500 range.

Then, I ask them, "Why pay so much?" Because they need graphics capability to play games. I then follow up with, "Can't you play games with a $50 graphics card?" The answer is no because the game would be so slow that it would be unplayable. My last question is what's the difference? Both cards can play the game, but one card is unacceptably slow compared with the other. The difference is that the faster card uses dedicated hardware to accelerate the game algorithm while the slower card must use software. In a nutshell, that's what partitioning is all about.

Once the product features are specified, partitioning takes over and is arguably the most important part of the design process because how the embedded system is partitioned will drive all the hardware and software developments to follow.

Of course, there are trade-offs that have to enter into any partitioning decision. Here's what we discuss in my class on microprocessor system design:

Advantages of a hardware solution
- Can be factors of 10×, 100×, or greater speed increase.
- Requires less processor complexity, so overall system is simpler.
- Less software design time required.
- Unless a hardware bug is catastrophic, workarounds might be doable in software.

Disadvantages of a hardware solution
- Additional HW cost to the bill of materials.
- Additional circuit complexity (power, board space, RFI).
- Potentially large nonrecoverable engineering (NRE) charges (~$100K+).
- Potentially long development cycle (3 months).
- Little or no margin for error.
- IP royalty charges.
- Hardware design tools can be very costly ($50–100K per seat).

Advantages of a software solution
- No additional impact on materials costs, power requirements, circuit complexity.
- Bugs are easily dealt with, even in the field.
- Software design tools are relatively inexpensive.
- Not sensitive to sales volumes.

Disadvantages of a software solution
- Relative performance versus hardware is generally very inferior.
- Additional algorithmic requirements may force more ○ processing power:
  - Bigger, faster, processor(s).
  - More memory.
  - Bigger power supply.
- RTOS may be necessary (royalties).
- More uncertainty in software development schedule.
- Performance goals may not be achievable in the time available.
- Larger software development team adds to development costs.

Of course, modern FPGAs have embedded cores, typically one or more ARM processor cores. In a way, this represents the best of both worlds as a partitioning environment. With the correct software development tools, it is entirely possible to design our embedded system starting from only one development environment and partitioning entirely within that environment.

If we think of an algorithm in a more generalized way, then we can see that you can think of software-only at one end of a scale and hardware-only at the other end of the scale. Along the scale is a slider that enables us to continuously change the partitioning of the design between the two possibilities.

This is possible because custom hardware generally implies an FPGA or custom ASIC. These devices are designed using hardware description languages such as Verilog or VHDL. Given that we are using software to design hardware, it isn't much of a stretch to make the leap to a single design methodology that incorporates simultaneous software and hardware design.

Phase 3: Iteration and implementation: Nane et al. [4] did a survey and evaluation of high-level synthesis tools (HLS) for FPGA development. High-level synthesis enables a designer to code an algorithm in C or C++ and the output of the compiler would be Verilog or VHDL. Fig. 6.4 is a reproduction of part of the table. The partitioning aspect of these tools is to enable the designer to code key algorithms without the need to consider how the algorithm would be implemented. Then, the next step is compiling the algorithm using the HSL tools and having the information necessary to make informed decisions about partitioning the design.

For no other reason than I'm familiar with Synopsis CAD tools, I looked into their Synphony C HLS tool. In a white paper (Eddington [5]), Synopsis discusses how the Synphony C compiler

Compiler	Owner	License	Input	Output	Year	Domain	TestBench	FP	FixP
eXCite	Y Explorations	Commercial	C	VHDL/Verilog	2001	All	Yes	No	Yes
CoDeveloper	Impulse Accelerated	Commercial	Impulse-C	VHDL Verilog	2003	Image Streaming	Yes	Yes	No
Catapult-C	Calypto Design Systems	Commercial	C/C++ SystemC	VHDL/Verilog SystemC	2004	All	Yes	No	Yes
Cynthesizer	FORTE	Commercial	SystemC	Verilog	2004	All	Yes	Yes	Yes
Bluespec	BlueSpec Inc.	Commercial	BSV	SystemVerilog	2007	All	No	No	No
CHC	Altium	Commercial	C subset	VHDL/Verilog	2008	All	No	Yes	Yes
CtoS	Cadence	Commercial	SystemC TLM/C++	Verilog SystemC	2008	All	Only cycle accurate	No	Yes
DK Design Suite	Mentor Graphics	Commercial	Handel-C	VHDL Verilog	2009	Streaming	No	No	Yes
GAUT	U. Bretagne	Academic	C/C++	VHDL	2010	DSP	Yes	No	Yes
MaxCompiler	Maxeler	Commercial	MaxJ	RTL	2010	DataFlow	No	Yes	No
ROCCC	Jacquard Comp.	Commercial	C subset	VHDL	2010	Streaming	No	Yes	No
Synphony C	Synopsys	Commercial	C/C++	VHDL/Verilog SystemC	2010	All	Yes	No	Yes
Cyber-WorkBench	NEC	Commercial	BDL	VHDL Verilog	2011	All	Cycle/ Formal	Yes	Yes
LegUp	U. Toronto	Academic	C	Verilog	2011	All	Yes	Yes	No
Bambu	PoliMi	Academic	C	Verilog	2012	All	Yes	Yes	No
DWARV	TU. Delft	Academic	C subset	VHDL	2012	All	Yes	Yes	Yes
VivadoHLS	Xilinx	Commercial	C/C++ SystemC	VHDL/Verilog SystemC	2013	All	Yes	Yes	Yes

Fig. 6.4 Currently available high-level synthesis tools. From R. Nane, V.-M. Sima, C. Pilato, J. Choi, B. Fort, A. Canis, Y.T. Chen, H. Hsiao, S. Brown, F. Ferrandi, J. Anderson, K. Bertels, A survey and evaluation of FPGA high-level synthesis tools, IEEE Trans. Comput. Aided Des. Integr. Circuits Syst. 35(10) (2016) 1591.

supports higher levels of abstraction and blurs the line between hardware and software. These are:

- Enables exploration from a single sequential C/C++ algorithm.
- Provides a balance of design abstraction and implementation direction to build efficient hardware.
- Easy partitioning of the programmable and nonprogrammable hardware.

Another feature of the Synphony C compiler is the inclusion of a tool called an Architectural Analyzer. The tool enables a user to import unmodified C/C++ code and then try possible optimizations and see the performance trade-offs that result.

However, these tools are not for everyone. They are complex and costly. It takes a dedicated investment in time and resources to master their use. However, if you can devote the time to purchase, learn, and use them, then defects that might be introduced during the partitioning process can be avoided.

This is critically important if the hardware will be an ASIC because of the high up-front cost of ASIC design and fabrication, but also because of the time involved in manufacturing a custom integrated circuit. For an FPGA-based design, this is not the case because it can be reprogrammed at any point in time. The disadvantages of the FPGA relative to the ASIC are apparent in several areas. Singh discusses these differences in an Internet article [6], and they may be summarized as follows:

- The FPGA is reconfigurable while the ASIC is permanent and cannot be changed.
- The barrier to entry for an FPGA design is very low while the cost and effort to do an ASIC is very high.
- The key advantage of the ASIC is that of cost in high volumes where the NRE and manufacturing costs can be amortized.
- The ASIC can be fine-tuned to lower the total power requirements while this is generally not possible with an FPGA.
- The ASIC can have a much higher operating frequency than an FPGA because the internal routing paths of the FPGA will limit the operating frequency.
- An FPGA can have only limited amounts of analog circuitry on the chip while an ASIC can have complete analog circuitry.
- The FPGA is the better choice for products that might need to be field upgraded.
- The FPGA is the ideal platform for prototyping and validating a design concept.
- The FPGA designer doesn't need to focus on all the design issues inherent in an ASIC design, so the designer can just focus on getting the functionality correct.

It is also interesting to note that in today's climate of hacking and cybersecurity, an FPGA can represent a security risk. If a malevolent entity such as a rogue government can gain access to a key part of a country's telecomm infrastructure, such as a data switch, then changing the FPGA code could easily be more difficult to detect than the same hacking in the software.

Let's consider debugging in the context of cybersecurity. When an embedded device is hacked, then we have a software or hardware defect in the product. This defect can be considered to be the vulnerability of the device to hacking, or it can be the hack itself. Many of the techniques we would use to find and fix a bug are the same whether you are trying to find a vulnerability in a device or whether the flaw is manifested as a bug in normal operation. So, while the use of an FPGA in released products brings many benefits, there is a class of infrastructure-critical devices that can be compromised in ways that are very hard to detect. Not impossible to detect, but very hard.

As the hardware and software teams begin to start their separate design processes, we can assume that the boundaries between the requirements, as championed by the stakeholders in marketing, the hardware team, and the software team, are still rather fluid. What the customer may need, or claim to need, versus the price point, competition, needed technology, and all the other factors is generally under constant scrutiny and debate. At some point, these issues must be resolved, and the features and boundaries are frozen and agreed upon.

We all are familiar with the term "feature creep." There is even a Dilbert cartoon with an ogre-like character called the Feature Creep whose sole function is to tell the engineers to add more features whenever they think they're done.

Feature creep results whenever:

- The product definition is weak.
- A marketing engineer just got back from visiting a big customer.
- The competition just introduced a new product that usurps your proposed new product.

The worst thing to do is panic, decide to put on a bandage and add some more features, then not adjust the project schedule. I recall a presentation by an R&D project manager at an HP Project Management Conference[e] on just this phenomenon. The manager's suggested "best practice" was that, at the start of every status update meeting, he would ask the assembled engineers and other stakeholders if anything has changed in terms of

[e] I was a member of the HP Project Management Council at the time and our group organized the conference.

features. If anything has changed, the manager declares that the project is on hold until the issue can be resolved, and a new schedule is generated. At the next meeting, the impact of the requested change is assessed, and the decision is made to add the feature and slip the schedule, or not.

By doing this, the manager forces all the stakeholders to do a reality check. If the feature is worth the risk, or the product would be noncompetitive without revising the specs, then a new schedule is in place and the new features are added to the schedule and to the design. Of course, this also has the effect of unfreezing the partitioning of HW and SW so the ripple effect can be quite significant. The key is that a feature adjustment after the initial partitioning is completed is an ever more serious perturbation to the project schedule and all the stakeholders need to assume ownership for decisions that might impact the development schedule.

Once again, we need to take the broad view of debugging during this phase of the project. Debugging here could involve running performance tests on the processor of interest. This could be accomplished using an evaluation board from the silicon manufacturer and some representative code that would mimic the actual loading on the processor.

The Embedded Microprocessor Benchmark Consortium (EEMBC)[f] is an organization made up of member companies within specific application disciplines such as automotive, office equipment, avionics, silicon manufacturers, and tool vendors.

According to their web site:

> *EEMBC benchmark suites are developed by working groups of our members who share an interest in developing clearly defined standards for measuring the performance and energy efficiency of embedded processor implementations, from IoT edge nodes to next-generation advanced driver-assistance systems.*
>
> *Once developed in a collaborative process, the benchmark suites are used by members to obtain performance measurements of their own devices and by licensees to compare the performance of various processor choices for a given application. Recently developed EEMBC benchmark suites are also used throughout the community of users as an analysis tool that shows the sensitivity of a platform to various design parameters.*[f]

[f]https://www.eembc.org/about/.

The need for benchmarks that mirrored the actual algorithms that a particular user segment would typically require became a necessity when compiler manufacturers realized that they could improve sales by optimizing their compilers for the most common benchmark in use at the time, the MIPS benchmark. The MIPS benchmark is actually derived from the Dhrystone benchmark and was referenced to the old Digital Equipment Corporation VAX 11/780 minicomputer. The 11/780 could run 1 million instructions per second, or 1 MIPS, and it could execute 1757 loops of the Dhrystone[g] benchmark in 1 s. The Dhrystone benchmark was a simple C program that compiled to approximately 2000 lines of assembly code and was independent of any O/S services. If your microprocessor could execute 1757 Dhrystone loops in 1 s, it was a 1 MIPS machine.

As soon as the compiler vendors started to tweak their compilers to optimize for the MIPS benchmark, the acronym changed to:

Meaningless indicator of performance for salesmen

The EEMBC consortium was driven by Marcus Levy, the technical editor of EDN magazine. He brought users and vendors together to form the core group. The first "E" in EEMBC originally stood for EDN but has since been dropped from the name of the organization. The first "E" was kept because the name had become so widespread.

The EEMBC benchmarks provide a consistent set of algorithms that can be used *for relative performance measurement.* Taken in isolation, a processor benchmark may not be very useful because it is subject to the compiler being used, the optimization level, the cache utilization, and whether the evaluation board accurately reflects the processor and memory system of the actual product. These evaluation boards were called "hot boards" because they were designed to run with the fastest clock and lowest latency memory.

Other factors that could negate the results included RTOS issues such as priority level and processor task utilization. However, at the very least, system architects and designers would have a significantly more relevant code suite to use to predict processor performance for a given application.

To demonstrate the effect of the interplay between hardware and software performance, and why the EEMBC benchmarks are so valuable, I show the graph in Fig. 6.5 to my classes [7].

[g]https://en.wikipedia.org/wiki/Dhrystone.

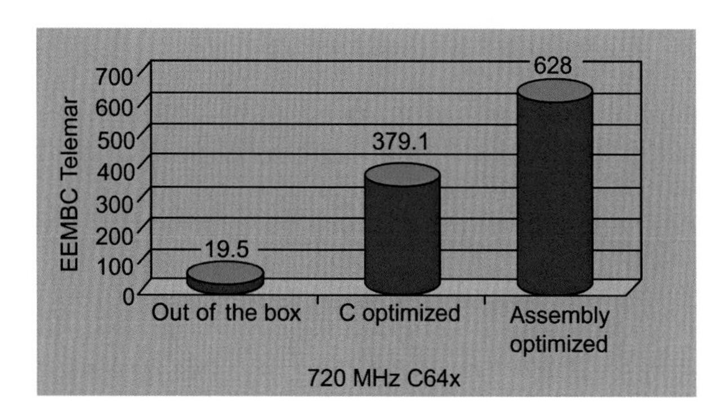

Fig. 6.5 Relative performance of the TMS320C64x DSP processor running the EEMBC Telemark benchmark. Courtesy of EEMBC.

The three columns represent EEMBC scores on the EEMBC Telemark Benchmark. This benchmark is one of a set of benchmarks that make up the TeleBench suite of benchmarks. This suite of benchmarks allows users to [8]:

> *approximate the performance of processors in modem and related fixed-telecom applications.*[h]

The leftmost column shows the benchmark score for the TMS320C64x DSP processor compiled without any optimization switches enabled, resulting in a benchmark score of 19.5. When various optimization strategies are employed, particularly those that can take advantage of the architecture of the processor, the benchmark score dramatically improves. Specifically, the TMS320C64x has 2 identical groups of 4 functional units and 2 identical banks of 32 32-bit general-purpose registers.

With the compiler able to aggressively take advantage of these architectural features, the benchmark score jumps to 379.1, which is an improvement of more than 19 times. In column 3, the code is hand optimized at the assembly language level, resulting in an improvement in the benchmark score by almost another factor of 2. The total improvement from out of the box to hand-crafted in assembly language is $32\times$.

To put this performance data in context, this is a potential bug that will only become apparent during validation testing, when the system loading begins to stress the processor's ability to

[h]https://www.eembc.org/benchmark/telecom_sl.php.

handle it. The scenario might be that deadline failures are observed during testing and the engineers begin the process of debugging the code. However, there is nothing wrong with the algorithms themselves, as the defect lies in the decision of how to compile the code.

It might have been something as simple as the need to turn off optimizations during testing, and the bug was not able to revise the make-file to turn the optimizations back on. That's certainly happened before.

The key takeaway here is that these performance issues need to be resolved sooner rather than later. In fact, all the compiler issues should be part of the internal specification document that will define the development environment of the project.

Phase 4: Detailed HW/SW Design: This is the phase that everyone is most familiar with because this is the phase where the hardware and software bugs are predominantly introduced into the project. However, I hope that I've at least sensitized you to the reality that defects can be introduced far earlier in the process due to poor decisions in processor selection or design partitioning. While clearly not the same as missing trace on a PCB, a project decision having to do with the poor choice of a vendor for a critical part can have the same level of schedule impact as trying to chase down an elusive hardware glitch.

My next favorite defect is the hardware bug workaround. This typically occurs later in the process when hardware and software are brought together for the first time. If the hardware is an FPGA, then it is generally a nonissue. If it is a custom ASIC, then it is a big problem. This is where the "fix it in software" solution is brought into the equation.

Now the advantage of hardware acceleration is lost and more of the burden of maintaining the desired level of performance is put upon the software team because the part of the hardware algorithm that does not work properly must be repaired/ replaced/augmented/etc. in software.

In the ideal case, software and hardware are incrementally integrated during the development process using the techniques described in earlier chapters. For example, as software modules are completed, there should be a test scaffolding available to exercise the module. At the lowest level of driver software that must directly manipulate the hardware, the necessity for early integration is most important. You want to catch errors as soon as possible, not have to find and fix the defects when the stakes are much greater and time is of the essence.

The same process of incremental integration is also critical for the hardware team. Again, in the ideal case, the low-level software drivers will be available to exercise the hardware, either for real hardware or through simulation techniques such as coverification or cosimulation. Here, the drivers are used to exercise the hardware while the ASIC is still HDL code.

Tracing defects in this phase is a lot simpler because the number of variables that can be the root cause of the problem is smaller and more manageable. Also, following good design processes such as running simulations and validating designs with formal design reviews are worth their weight in gold, and will filter out many potential defects before they make it to production.

Glitches are the bugs that give us nightmares. A glitch is usually so infrequent that we would be lucky if we detected it during this phase. A software glitch, such as a priority inversion within the RTOS or a stack overflow, might not show up until the system is fully loaded and operating under actual conditions, rather than unit testing. In my opinion, a hardware glitch is even more of a challenge because of the number of possibilities.

My first introduction to hardware glitches occurred while I was a graduate student. We were using a high-voltage pulse circuit that used a mercury-wetted relay to generate a 0–5 kV pulse with a 1 ns rise time. Located in the same room as the pulse generator was our minicomputer (I hope this doesn't date me too badly) that controlled the experiment. A bundle of cables exited the minicomputer rack and encircled the lab, going to various sensors and detectors. Everything worked properly until we fired the pulse generator while the minicomputer was logging data from some of the more remote sensors. Then the program would crash.

Because graduate students have an infinite amount of time for their thesis and with no endpoint in sight, I set about trying to figure out what was going on. I'll spare you the details of how I eventually found the source of the problem and cut right to the chase. The source of the problem was the radiated energy from the pulse generator whenever the pulse amplitude got above about 2 kV. This was picked up by the ground shields on the various cables, and they were transmitting enough energy back to the minicomputer power supply that we could see a ground bounce of several volts.

This is the part where I learned about optical isolation. I rebuilt the data acquisition hardware to isolate the data logging and signal conditioning from the digital I/O to the computer. The minicomputer ground was then isolated in the equipment rack. Problem solved.

Detecting glitches is what keeps oscilloscope and logic analyzer manufacturers in business.[i] Manufacturers such as Tektronix and Keysight provide a wealth of application and instructional data for students and engineers alike. While researching glitches, I came upon an interesting Keysight video[j] that discusses how to use the fast Fourier transform (FFT) capability built into their oscilloscopes to detect potential sources of glitches. I thought this was rather clever, so I watched the video and learned that in modern high-speed digital systems, crosstalk between components is becoming an ever more common problem. The video showed a clock signal with some high-frequency noise riding on it. I would have normally dismissed this as ground bounce, but when the FFT was enabled, it showed that the high-frequency noise was at 19 MHz and was due to crosstalk in the circuit. Crosstalk can lower noise thresholds and make infrequent glitches more likely. My approach would likely have been to stay in the time domain and try to set up the oscilloscope for a single capture glitch detection trace.

When a glitch occurs inside an FPGA, then detecting the glitch becomes more challenging. One method is to incorporate a glitch detection circuit within the FPGA along with the hardware algorithm being implemented [9]. The article points out that simulation can only go so far because whether a glitch will or will not occur in a particular routing of the FPGA may also be dependent upon the FPGA implementation. Also, assertions within Verilog are common ways to detect glitches if they occur.

A digital glitch detector circuit is described in a US patent for a logic analyzer [10] that appears easy to implement in an HDL and can detect either a positive or negative going glitch within a clock cycle. Thus, during testing of your FPGA circuit, this glitch detector can be added to the FPGA with very little extra resources required and provide a registered output signal that a glitch has occurred on the data. This type of detector can easily be added to the design and removed once the design is verified.

Suffice it to say that when you suspect that you have a glitch in the circuit, either because you see it (rarely) or you see its effect upon the circuit (most likely), the best approach is to start taking copious notes and head for the Internet. Students and experienced engineers alike can benefit from the infinite resource that is the Internet. A few hours of directed research can provide a wealth of information and insights into what could be causing the glitch and techniques for finding it.

[i] Just kidding.

[j] https://www.youtube.com/watch?v=u–OsHMGMYc.

The takeaway from this part of the embedded system development life cycle is that integration of hardware, software, test software, turn-on software, and anything else that will be needed when the final hardware and software are brought together should not occur in one Big Bang. As much as possible, it should be an incremental process with notetaking along the way so that when something goes awry, and we know that it will, you can trace back to what you know about the problem. While engineers are loath to document, and I freely admit to that rap, keeping a paper trail will ultimately save time in the end.

Perhaps it takes more time on the current project, but if you get in the habit and make ongoing documentation part of your process, then Murphy's Law guarantees that you will need it in the future.

Phase 5: HW/SW Integration: HW/SW integration was the traditional debugging phase. This is where tool vendors like me focused our product offerings and debugging solutions. In the classic model, a model that sold a lot of in-circuit emulators and logic analyzers, untested software meets untested hardware and may the best team win. The classic model also describes how the hardware team throws the embedded system "over the wall" to the software team and as soon as they hear the board hit the floor, they move on to the next project and are no longer available to the software team. The imagery is powerful, and it makes a good story, especially in the slide deck of a good sales engineer.

I think we're better at it today then we were 20 years ago, but I don't have any real data to back me up there. I think we've learned a lot about effective development processes, and, as I've pointed out, there are tools today that enable HW/SW integration to occur earlier in the embedded life cycle and in a more incremental fashion. Let's look at the HW/SW integration problem in some depth and see how debugging fits into it.

For this discussion, we'll assume that the only software of interest are the low-level drivers and the board support package (BSP) that an RTOS vendor, or the development team, must create if an RTOS will be used. The key challenge is reducing the number of variables that are in play when untested hardware meets untested software. Therefore, the debugging strategy should be based on eliminating as many of these key variables as possible so that the rest of the system integration becomes more tractable.

Step #1: Processor to memory interface: Whether the memory system is internal, external, or mixed, the interface between the processor core and the memory must be stable, or this is as far as you get. Memory decoding must be properly configured to identify memory regions, wait states, timing margins measured and recorded, etc. If the memory is static RAM, this process is relatively straightforward. For dynamic RAM, the challenges are greater.

Traditionally, this was the real power of the in-circuit emulator (ICE) because the ICE could still function even if the target system memory interface was not functioning properly. The reason is the ICE processor could run out of local memory, so a test program could run on the ICE and perform read and write tests to memory on the target board. This way, you could write a tight loop to read and write to various memory locations and observe the signal fidelity and timing margin using an oscilloscope.

Perhaps its unthinkable to you to imagine that the hardware team would turn over a board in such an untested state to the software team, so perhaps I'm being overly dramatic here. In any case, the memory interface is one of the first tests that needs to be performed.

Step #2: Programming the hardware registers: Properly initializing the hardware registers, whether on the processor microcontroller or the custom devices, can be a black hole for time. There is a semiconductor company that will remain nameless that would double or triple the functionality of the on-chip register set. Deciphering exactly how to properly initialize the registers from the hardware manual can drive an engineer to tears. This is especially true of a new chip, where one of the variables is the typo density of the user's manual.

One of the first companies to address this problem was Aisys. I crossed paths with them when I was responsible for third-party development tool support for AMD embedded processors. To the best of my knowledge, Aisys is no longer in business but their premier product, Driveway, was a software tool that would automatically create the driver code for popular microcontrollers at the time based upon a graphical and table-driven input specification.

Driveway was very expensive (more than $20K per seat), but their value proposition was the time they saved the product development cycle, reducing the time it took to write and debug the driver software from months to days or weeks.

Similar products are available (for free) today. For example, the Peripheral Driver Generator from Renesas Electronics is a free download that can be used to generate driver code. To quote Renesas [11]:

> *The Peripheral Driver Generator is a utility that assists a product developer in creating various built-in peripheral I/O drivers of a*

microcomputer and the routines (functions) to initialize those drivers by eliminating the developer having to do manual coding. All the necessary source codes are prepared by the Peripheral Driver Generator according to user settings, so that the development time and development cost can be greatly reduced.

Other semiconductor companies have similar offerings. NXP semiconductors offers SPIGen, a free SPI bus code generator that can adapt to a wide variety of SPI protocol specifications [12].

These tools take much of the headache out of creating the initialization and driver code for a wide array of embedded microcontrollers. Another factor that should not be underestimated is the sheer volume of code examples of every possible description available on the Internet. I think the most significant thing my students learn from taking my microprocessor class is how to find code and application examples online. They can't believe that I actually encourage them to use the code they find on the web, just so long as they cite their sources and give credit for assistance they receive.

Getting the driver code just right is one of the more challenging aspects of the HW/SW integration process. Just having one bit set wrong in one of the configuration registers can prevent the system from functioning, leaving you to wonder if you have a hardware or software fault to deal with. Actually, you're correct, because it is a hardware fault and a software fault. For example, the NXP ColdFire MFC5206e microcontroller has 108 peripheral registers, controlling all the I/O and memory bus functionality of the device [12]. A one-bit typo in one of the subfields of a register will easily cause the memory processor interface to fail. Fortunately, this microcontroller has extensive on-chip debug support (see next chapter), enabling the designers to debug the processor without a functional memory system. Using the on-chip debug resources, the memory-mapped peripheral control registers can be read and modified.

With on-chip control of the CPU through the debug resources, it is a straightforward process to load the turn-on and test code necessary to check the external memory interface, measure signal fidelity and bus timing, and generally verify that the hardware is ready to start taking the rest of the driver software, followed by the application software. If an RTOS is going to be used, then the board support package (BSP) would be installed at this point. The BSP drivers might make use of, or replace, the low-level drivers that you would need to create when not using an RTOS.

The key to success in the HW/SW integration phase is to proceed in deliberate steps, keeping notes along the way. Start from the most basic assumption that nothing works, other than the fact that the board probably won't catch fire when powered up (although this might be an erroneous assumption) and then start testing from the most basic to the more complex, always keeping notes and having a checklist of what tests to run next. Remember, your task is to reduce the number of possible variables that can be causing bugs to some manageable number. This is the key challenge of embedded systems with initial turn on of hardware and software.

Again, in the ideal case, the hardware team has already tested the hardware to the extent that they feel confident that the system is ready for the application software. If there are remaining hardware bugs, they'll be the corner cases that weren't tested for or missed communications between the teams, leading to errors in the device drivers. Also, marginal timing issues won't surface here because the hardware is being tested in a room temperature environment. It won't be until the system is subjected to thermal, humidity, and mechanical stresses that other hardware weaknesses become visible. Also, it won't be until the next phase when the system validation tests are being run that radio frequency (RF) testing will uncover out-of-compliance RF emissions and possible random errors due to crosstalk. Lucky you.

Phase 6: Acceptance testing and validation: Full disclosure. I hated this phase of the development cycle. I couldn't wait for it to be over. Environmental testing was the worst. We called it "shake and bake." The product was put on a vibration table and vibrations were introduced until resonance was hit. A strobe lamp was synchronized to the vibration table's frequency so you could see the components being distorted. I hated it. It was like watching my child being tortured.

Then came temperature and humidity cycling. This played havoc with any high-voltage circuitry[k] and we could hear the arcing in the chassis. It sounded like someone was cracking a whip. These tests were internal to HP and represented our validation of the robustness of the design.

I once asked the HP compliance engineer why they took the temperature up to 100°C because no lab instrument would ever get that hot. The reason was that was the benchmark temperature for the interior of the trunk of a dark blue car in Phoenix in the summer. The HP instrument didn't have to run at that temperature, only survive the heat exposure sitting in the truck of a field sales engineer's car. It was a good thing that we tested

[k]Remember, oscilloscopes had cathode ray tubes and there was 20 kV on the screen.

it because the plastic front grill on our ICE unit sagged from the heat and we had to use a different formulation for the plastic. Ditto for the cold temperature cycle, only this is Alaska in the winter.

The other series of tests were necessary for compliance with standards agencies such as the FCC and UL, and for the compliance agencies in Europe and Asia. For example, in Germany it is the TUV Rheinland that is responsible for electromagnetic compliance testing. Germany is an interesting situation because large factories would be located in small towns with residential housing butted right up to the building. RF interference could easily override broadcast TV and radio. TUV cars with rotating antennas would drive through the town and measure the RF emissions as they drove. If they detected RF amplitudes over the legal limit, a factory could be shut down.

While this really isn't the appropriate section to discuss RF issues, it probably is as good as any and even though this book is ostensibly about debugging, I would like to include this section about some best practices for RF design that I learned over the years. This is not intended to be a complete treatise on RF design techniques, but just a few thoughts that are easy to swallow and be sensitized to. Also, I already had a few slides on it because I teach these in my microprocessor design class, so they are easy to include.

Earlier, I talked about timing margins, but I didn't talk much about clock speed. Both are relevant here because they are tied up with RF issues. In embedded design, the general rule is to run the clock as slowly as possible and still accomplish the task at hand. In a PC, clock speed is a marketing tool, the faster the better. Overclocking, anyone?

The slower the clock, the less power is consumed. This is due to the CMOS technology that is used in modern microdevices. Also, slower parts are less costly than faster parts. With respect to RF, there are two related effects. The higher the clock speed, the greater the energy that is in the harmonics of the waveform. For a good square wave clock, it is easy to have energy out to the fifth harmonic.

The rising and falling edges also have an effect upon RF for other logic besides the clock. If you are using medium-speed devices, and you come up on timing issues where the worst-case propagation delay in one part violates the minimum set-up time requirement of another part, it might be tempting to just replace the offending part with a faster part. The faster part will generally have a short rise and fall time, and a faster rise time means more harmonics.

An article in EDN magazine [13] provides this simple rule of thumb that relates the rise time of a square wave to the effective bandwidth of the signal.

The bandwidth, in GHz = 0.35/Rise Time, in nanoseconds.

The article goes on to say,

Bandwidth is the highest sine wave frequency component that is significant in a signal. Because of the vagueness of the term "significant," unless detailed qualifiers are added, the concept of bandwidth is only approximate.

Bandwidth is a figure of merit of a signal to give us a rough feel for the highest sine wave frequency component that might be in the signal. This would help guide us to identify the bandwidth of a measurement instrument needed to measure it, or the bandwidth of an interconnect needed to transport.

From the RF perspective, bandwidth tells us about the RF frequencies that we will have to deal with and manage.

Therefore, switch to the fast logic with the knowledge that you are tempting fate, or at least Murphy's Law. Anyway, here is a list, in no particular order of precedence, of some of my general rules for good RF design practices:

- Use spread-spectrum clock oscillators to spread the RF energy out over a range of frequencies [14].
- Avoid current loops.
- Shield the clock lines on inner layers of the PC board, or run parallel guard traces.
- Avoid long clock lines.
- Avoid long bus runs on a board.
- Avoid using logic with fast logic edges: the ALS family is preferable to the FCT family if propagation delays are acceptable.
- Use RF suppression (ferrite) cores on cables.
- Shield locally, rather than the entire chassis.
- Run at the slowest acceptable clock speed.
- Terminate long traces in their characteristic impedance.

A few comments are in order here. It is generally much more cost effective to shield signals at the source rather than to have to come back after the fact and figure out how to shield an entire chassis. How long should a trace be before you need to add termination? We discussed this a bit in an earlier chapter, but it might surprise you to learn how long is "a long trace." For a signal with a rise time of 500 ps, the longest unterminated trace should be less than approximately 1.67 in. [15]. Therefore, unterminated traces reduce the noise immunity and generate crosstalk and RF energy that needs to be suppressed.

Most of the time we ignore the terminating signal in digital systems because we have a fairly wider noise margin than analog folks. However, depending upon the inherent advantages of digital systems to overcome poor electronic design practices is just asking for trouble.

What happens when we discover a hardware defect at this stage of the process? PCB issues are relatively straightforward to deal with. You fix the bug and manufacture new boards. Sometimes, if the fix is small enough, you do some rework of the board. The manufacturing folks really dislike this solution, but when the time crunch hits, this may be the only solution. Various companies had various policies about reworking PC boards. One of my former employers had the "five green wires" rule. More than five pieces of rework and you did a new PC board. Of course, we were talking about small volumes, less than 100 units per month. I wouldn't expect this rule to be too popular at mainstream electronics manufacturers.

If the hardware bug was in an FPGA, then fixing the defect is usually no more difficult than fixing a software bug. If the bug is in an ASIC, then fixing the bug could be much more involved. This is when the entire design could possibly unravel because the first thought will be, "Ok, just fix it in software. Do a workaround." But the very reason we have hardware is to accelerate the algorithm and reduce the demands on the processor.

This is going to take us right back to partitioning decisions. If our system design strategy is to depend upon custom hardware to do the heavy lifting and the microprocessor to do the communications and housekeeping, then a flaw in the ASIC may be impossible to fix in software without greatly crippling overall system performance. Conversely, if our design strategy is to wring the last little bit of performance out of the software, to the extent that we are hand-crafting the compiler output in assembly language, then we would be less likely to have to deal with a hardware fix because the hardware is not the critical part of the equation.

This is also the phase where we are stress testing the product and if it will be in a mission-critical application, testing it to the proper compliance levels that the certifying agencies require (FAA, FDA). For example, one of the most well-known requirements documents is DO-178C, *Software Considerations in Airborne Systems and Equipment Certification* [16].

Applied Microsystems Corporation, AMC, developed a software analysis tool called CodeTEST that contained built-in test suites and report generators for certifying software to the requirements of the certifying agencies, DO-178B[1] being one example.

I was responsible for the CodeTEST product line for a while during a 4-year stint at AMC. I left AMC to go into academia at the University of Washington just months before the company shut its doors and dissolved. CodeTEST was sold to Metrowerks, the software tools company. Metrowerks was then acquired by Motorola, who then spun off the semiconductor business to Freescale, which later somehow became NXP. Whew! The CodeTEST product line got lost in the shuffle around 2003–04.

CodeTEST was a combination hardware and software profiling tool. The software was used to preprocess the software for profiling, then the hardware tool was used to collect the profiling data in real time and with very minimal code intrusion. Then, the software took over again, postprocessed the data, and put it in the proper format for analysis or for certification.

The preprocessing involved placing "tags" at various locations in the code, such as function entry and exit points or program branches. These tags were simple "data writes" to a specific memory location or block of memory that was allocated to the CodeTEST hardware tool. The value of the data and the memory address provided the required information about where the tag came from. All the tags were time-stamped, and the data were buffered in the CodeTEST hardware and then sent to the host computer in burst packets when possible.

If you are having a case of déjà vu, don't fret. I have discussed this technology twice before in this book. I mentioned it as a performance measurement technique using logic analyzers and as an accessory tool for HP emulators. CodeTEST differed because it attempted to supply a total solution, rather than pieces, the way the other products did. The preprocessing was transparent to the software developers because the magic occurred in the "makefile" where all compilation and linking was invoked to build the software image. The CodeTEST preprocessor was invoked here to add the appropriate tags to the source code before compilation took place.

With respect to compliance testing, the biggest advantage of CodeTEST was its ability to show how well the validation software testing was actually meeting the requirements for code certification. Code coverage was one of the most difficult requirements to meet because there are so many possible code paths in a program of any reasonable complexity. In fact, there have been various

[1]DO-178B was later replaced by DO-178C.

statistical calculations that show the number of distinct paths through a program is greater than the number of stars in the known universe. This sobering fact makes it a real challenge to design test software that will prove to the FAA that there are no dead spots lurking in the code that have never been tested and will pop up at the most inopportune times.[m]

The HP 67000 family had an extra bit in its emulation memory that was put there for code coverage measurements. Every time that memory location was accessed (hit), the bit was set. The number of set bits could be counted, and we would be able to easily figure out how well the test code was actually testing the product. We actually used this ourselves to test our emulators and it was surprising how low our coverage numbers were when we first started testing.

As I recall (please don't quote me on this), our requirement was 85% coverage and initial tests usually resulted in 40% coverage.

In addition to using automated tools, we depended upon various forms of black box and gray box tests. "Abuse testing," as it was known, was something every engineer had to do and was written into our schedules. We typically were flagged to test someone else's product, not our own, because we knew where the warts were. The idea was simple: break the code so the product would freeze up or do something wrong. When a bug occurred through abuse testing, it was categorized at various levels. The highest levels were "critical" and "serious." When those bugs occurred, the testing stopped, and the bug report was sent to the designer to be fixed. This reset the clock and testing started all over again from the beginning. In order to release the product, there had to be no serious or critical defects in some number of hours (for the sake of argument, 10 hours).

The abuse testing was augmented by keystroke recording so that the engineer didn't have to restart testing by hand. The keystrokes were played back to the point of the original failure and the engineer took off from there. My particular favorite was the "falling asleep at the keyboard" test where I would put my head down on the keyboard and just let keys autorepeat for 5 or 10 min while I took a quick nap.

Phase 7: Product release, maintenance, and upgrade: The product has been signed off for release, marketing and sales are wound up and ready to go. All the swag has been purchased with the company logo, ready for the next conference and now, the most

[m] There is an embedded systems legend that I heard but cannot find a citation for about the F-16, the first all fly-by-wire fighter plane, that suddenly flipped upside down when it flew across the equator.

important person in the company takes over. "Who is that" you ask? According to this consultant at a seminar I attended, the most important person in the company is the hourly employee in the shipping department or on the loading dock who puts the boxes in the delivery van on the first step in the new product's journey to your customer.

Now begins the real defect testing. Everything up to now has been sterile and controlled, but now, the masses will take over and they will do things that were never imagined by the designers. Today, with FLASH memory holding the operational code, bug fixing is simply a software download. I think we're all familiar with this process. I also suspect that our ability to fix bugs in the field has mitigated the need to find and fix all the bugs in the factory. Has this made us sloppier? I don't know. However, it has made upgrading easier. Remember all those features that we wanted to add during the specification phase but were ruled out? With customer feedback and social media, we have our market research just one click away to tell us what we need to fix and add to the product. No more focus groups needed. Of course, this is a gross oversimplification, but modern technology has certainly changed how we deliver "the whole product" to the customers. Typos in user manuals are easily fixed, and the PDF files of the manuals are updated on the web site. Paper manuals shipped with the product are a thing of the past.

Wrapping up this chapter, I think the key message here is that the process of integrating embedded software with the hardware should be an incremental process, rather than a major event at the back end of the development cycle. There are tools and processes available today that make incremental integration a straightforward process.

While I'm writing this chapter as if the reader is the design engineer, my real target is the students who will be entering the field in a year or less. Real engineering is not accomplished by pulling a few all-nighters before the project is due. Well, maybe sometimes we need to pull an all-nighter, but in any sane organization, that is the exception, not the rule. For the student who is out there interviewing for her first EE job, what is going to sell you to the company and make you stand above the other job seekers is how well you can present yourself as ready to step in and be productive from the first day in the R&D lab.

Imagine that you are being interviewed by an R&D manager and you are asked to discuss your senior project. Your response,

> Well, my group was tasked with designing an autoranging LCR meter. We started by surveying the existing products in the market and researching the available technologies. We next

flushed out the feature set we wanted to have and that we thought we could achieve in the time we had available.
Our next task was to partition the design into hardware and software and map out the major functional blocks and interfaces between these blocks. A big part of this initial design was to choose the right processor and software tools as well as search for as many design examples as possible. I also visited a local engineering company and showed them our front-panel mock-ups and got their feedback.
Once we were satisfied with our starting point, we developed the specification documentation, the test plan, the validation plan, and the initial project schedule. As we begin developing the hardware and software, we had to make some changes to our specs, which then impacted the partitioning and the schedule, but we froze the design shortly after that.
Our team did periodic code inspections, trading off with other teams, and we had a formal hardware design review before releasing the board to fabrication. We also wrote test software to simulate the hardware and the software was exercised against the hardware APIs until we had real hardware to test on.
Our LCR meter worked as designed. We came within 3 days of our scheduled completion date. The PCB required one patch, due to a typo in the data sheet for the LCD display we bought.
Oh, by the way, here it is. We did a three-dimensional printing of the case. It will run for a year on three AAA batteries.

References

[1] E.F. McQuarrie, Customer Visits: Building a Better Market Focus, third ed., M.E. Sharpe, London, 2008. ISBN: 978-0-7656-2224-2, 2008.
[2] http://www.hpmuseum.net/display_item.php?hw=219.
[3] E. Wilson, Product Definition Factors for Successful Designs (M.E. Thesis), Stanford University, December 1990.
[4] R. Nane, V.-M. Sima, C. Pilato, J. Choi, B. Fort, A. Canis, Y.T. Chen, H. Hsiao, S. Brown, F. Ferrandi, J. Anderson, K. Bertels, A survey and evaluation of FPGA high-level synthesis tools, IEEE Trans. Comput. Aided Des. Integr. Circuits Syst. 35 (10) (2016) 1591.
[5] C. Eddington, C/C++ for Complex Hardware Design, A White Paper, November, https://www.synopsys.com/cgi-bin/proto/pdfdla/docsdl/cplus_chd_wp.pdf?file=cplus_chd_wp.pdf, 2010.
[6] S. Rohit, FPGA Vs ASIC: Differences Between Them and Which One to Use? https://numato.com/blog/differences-between-fpga-and-asics/, 2018.
[7] J. Brenner, M. Levy, Code efficiency and compiler directed feedback, Dobb's J. 355 (2003) 59. Now available on the web: http://www.drdobbs.com/code-efficiency-compiler-directed-feedb/184405506.
[8] https://www.eembc.org/benchmark/telecom_sl.php.

[9] R. Velegalati, K. Shah, J.-P. Kaps, R. Velegalati, K. Shah, J.-P. Kaps, Glitch detection in hardware implementations on FPGAs using delay based sampling techniques, in: Proceedings of the 16th Euromicro conference on digital system design, DSD 2013, 2013, pp. 947–954.
[10] K.A. Taylor, Glitch Detector, US Patent #4,353,032, 1982.
[11. https://www.renesas.com/eu/en/products/software-tools/tools/code-generator/peripheral-driver-generator.html.
[12] https://www.nxp.com/docs/en/data-sheet/MCF5206EUM.pdf, Appendix A.
[13] E. Bogatin, https://www.edn.com/electronics-blogs/bogatin-s-rules-of-thumb/4424573/Rule-of-Thumb–1–The-bandwidth-of-a-signal-from-its-rise-time, 2013.
[14] http://www.maxim-ic.com/app-notes/index.mvp/id/1995.
[15] http://www.interfacebus.com/Design_Termination.html#b.
[16] L. Rierson, Developing Safety-Critical Software: A Practical Guide for Aviation Software and DO-178C Compliance, CRC Press, Boca Raton, FL, ISBN: 9781439813683, 2013, p.198.

On-chip debugging resources

Introduction

It is hard to imagine debugging an embedded application without resources built into the processor core. Today, we take that for granted. In the early days of embedded applications, that wasn't the case. Every transistor was precious, as was every I/O pin on a package. The thought of adding extra transistors to support the development process and then shipping these processors to customers with the overhead of unused circuitry was unthinkable. Times have changed.

Modern process technology and hardware description languages, such as Verilog and VHDL, have basically made the cost of additional on-chip debug circuitry irrelevant, and any modern microcontroller has the expectation of a rich set of on-chip debug resources. Throwing a debug core onto a very basic microcontroller adds almost nothing to the cost per die and has become a customer requirement.

It is interesting to note that the evolution of the on-chip debug core was the technological advancement that led to the demise of the in-circuit emulator as the premier tool for embedded systems design and debug. Emulator companies took pride in the ability of their ICE boxes[a] to transparently control and become the processor being emulated through very sophisticated external circuitry that took over access to the nonmaskable interrupt input of the processor. Once you add a debug core to the chip that performs all the functionality (and more) of the external circuitry, and does it for free, the value proposition of the ICE box quickly vanishes.

While in its most primitive form, on-chip debugging resources are no more feature-rich than a traditional software debugger, there is one key advantage that makes the on-chip debug core so valuable. Simply put, it doesn't depend on the processor-to-memory connection. The debugger does not reside in memory,

[a]ICE is the acronym for in-circuit emulator.

Debugging Embedded and Real-Time Systems. https://doi.org/10.1016/B978-0-12-817811-9.00007-7

nor does it depend upon a stable processor-memory interface. For processors with small memory footprints, this is a blessing because if you only have 2 K bytes of program space, dedicating 1 K to the debug kernel is a significant drain on memory capacity.

This was one of the primary value propositions of the ICE box, so it is easy to see why the on-chip debug technology so dramatically impacted emulator sales.

On-chip resources for debugging have been around in various forms for quite a while. The first microcontroller that I am aware of that had on-chip resources for debugging was the 8051 family from Intel. At one point in time, the 8051 and its derivatives were the most popular microcontrollers in existence. I don't the exact numbers here in terms of percentage, but I recall seeing some early marketing data that the 8051 family owned more than 75% of the market for embedded microcontrollers. In an oral history panel discussion held at the Computer History Museum in September 2008 [1], seven of the key engineers who developed the 8051 were interviewed. Of interest in this context was that one of the first versions of the chip that was developed was the bond-out version, designed to support ICE development, particularly the group at Intel that designed the ICE units for their processors.

At the time, sales of in-circuit emulators represented a profit center for a silicon manufacturer, and Intel guarded the technology of the bond-out chip so that only Intel emulators[b] could transparently emulate the 8051. Without going into too much detail because I'm not sure where the time limit on my proprietary knowledge ends, I will say that in the mid-1990s, there was a huge and costly legal battle between AMD and Intel over AMD's license to second-source Intel microprocessors. At the heart of this legal battle was the 8051 and the bond-out technology.

The bond-out version of the chip had additional I/O pins that provided external circuitry with important information about what was happening internally because the address, data, and status buses were invisible to the outside world. The bond-out chip also facilitated setting break points for code development. If we roughly assume that the 8051 first became widely available in the 1980s, then we can tie the first use of on-chip debug resources to that timeframe.

Motorola was also an active player in the embedded marketplace and in a similar timeframe as Intel developed the extremely popular 683XX family of microcontrollers. These chips, and their

[b]Affectionately known as "Blue Boxes" because of the chassis color.

follow-ons, contained Motorola's version of on-chip debug, called Background Debug Mode, or BDM [2].

Background debug mode

The on-chip BDM resources were consistently found on a wide variety of Motorola[c] embedded controllers up from the original CPU16 and CPU32 devices to their ColdFire family and later the PowerPC families. The interface was consistently similar, except for minor differences due to architectural differences in the processors. The exception is the PowerPC, which used a different architecture for its BDM implementation.

BDM was implemented as a 16-bit serial bit stream and a 17th status/control bit. The typical interconnection standard was a 10 or 26 connector. The basic 10-pin connector provided the standard debug functions that we would expect to see in a software debugger and the 26-pin interface added real-time trace capabilities to the interface.

The basic BDM command set can be summarized as shown below:

Read register	RAREG/ RDREG	Read the selected address or data register and return the result
Write register	WAREG/ WDREG	Write the specified value to the selected address or data register
Read memory	READ	Read from the specified memory location
Write memory	WRITE	Write to the specified memory location
Memory dump	DUMP	Read from a block of memory
Memory fill	FILL	Write to a block of memory
Resume execution	GO	Resume instruction execution at the current value of the PC (after pipeline flush)

These basic commands might also include one or more extension words that provide other data values associated with the command. The data into the processor core is shifted along the serial-data-in (DSI) pin and the output data is shifted out on the serial-data-out (DSO) pin. The data transfer clock (DSCLK) is provided by the external debugger.

[c]Motorola's semiconductor business was spun off and became Freescale, which later sold off the processor business to NXP.

The interesting fact that BDM could be implemented as a 10- or 26-pin interface was later adopted by the IEEE 5001 NEXUS standard,[d] which defined a scalable debug architecture along the same lines. We'll discuss this standard later in the chapter.

For example, with the 26-pin interface, there are 8 additional signals that can output in real time, including 4-bit wide information about the processor's state. If your debugger has some logging capability, general knowledge of the code being executed, and a linker output map along with some clever host software, much of the processor's real-time program execution can be reconstructed.

Unlike the proprietary bond-out parts that Intel built for use by its internal team building ICE units, the BDM standard was public and could be accessed by any tool vendor who wanted to build hardware and software tools to support Motorola microcontrollers.

JTAG

The basic JTAG[e] standard, IEEE 1149.1, defines a methodology of designing for testability, DFT. It came out of the automated test Industry and was driven by the need for a better way to test complex printed circuit boards. Before JTAG, a large computer board would need to have a dedicated engineer design a test fixture to test the board. These test fixtures were mounted in a large, complicated, and expensive testing device called a "bed of nails" tester, shown below in Fig. 7.1.

The bed of nails is the array of spring-loaded, gold-plated pointed probes. Each probe comes up from below the board in the fixture and contacts a circuit node on the board. By making contact with every node in the system, each "nail" can either monitor the voltages on the node, or drive a signal into the node with a voltage or current.

It is easy to see how a PC board with hundreds or thousands of nodes could easily require hundreds of engineering hours to build the fixture, set up the bed of nails, program the tester, and so forth.

The development of JTAG enabled board testing to be built into the board by way of the components themselves. The name JTAG comes from the industry group that formed to codify the standard, but the technical term for the methodology is called "boundary scan" [3].

[d] https://nexus5001.org.

[e] JTAG = Joint Test Action Group.

Fig. 7.1 Bed of nails test fixture. Courtesy of SPEA spa.

Boundary scan requires that every device I/O pin include a register that can be read or written to. All these registers are connected in one long, continuous scan path. Sort of a serial bit stream on steroids. There can easily be thousands of bits in the stream, depending upon the total number of nodes in the system. This is shown in Fig. 7.2.

Not shown in Fig. 7.2 is the serial clock input necessary to synchronize the bit transfers.

It didn't take IC designers very long to realize that what could be done for a board test could also be used for microcontroller debugging. The idea was to link all the internal registers and other important circuit blocks in a JTAG loop that could both sample a register and modify it as well.

With the JTAG loop, additional debug-only registers could be easily added and accessed along with the CPU's general register set. What results is an on-chip debug capability that is built around an IEEE standard rather than a proprietary interface.

The introduction of the JTAG debug interface as a low-cost debug standard was a great leap forward for silicon manufacturers and end users, but not so great for tool vendors, specifically ICE manufacturers.

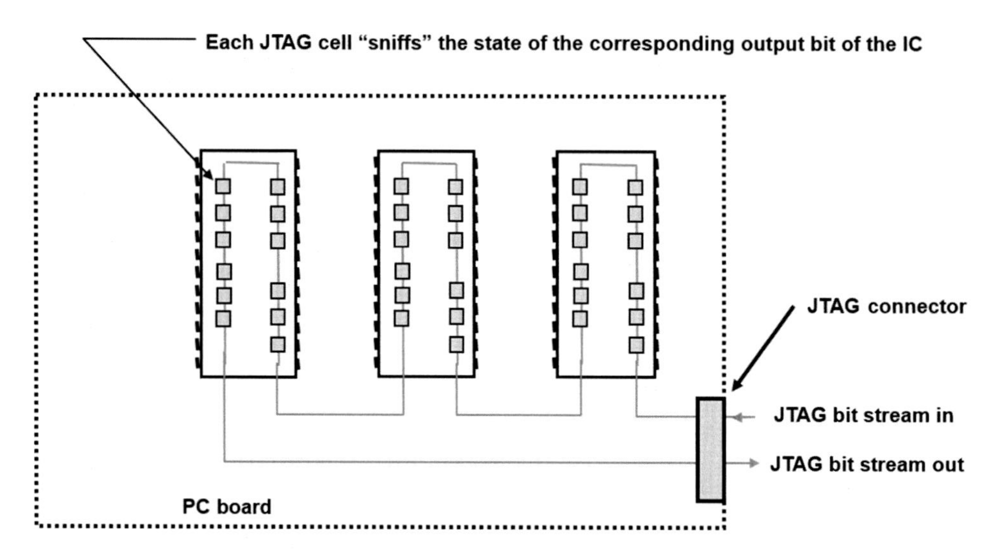

Fig. 7.2 Schematic representation of a boundary-scan testing loop.

Before we delve deeper into the features of these on-chip debug resources, let's look at one more standard that I was involved in for several years. This is the IEEE 5001 Nexus Forum standard.[f]

The standard was driven by the automobile industry and grew out of a plea to their semiconductor and development tool vendors to offer one interface to processors and to development tools so that the customers (the auto industry) did not have to buy and learn new tools every time they switched their processors in their automotive products.

To quote a white paper issued by the Nexus Forum [4],

> *The goal of the Nexus standard is deceptively simple: to provide a standard way for programmers and engineers to see what's going on inside their systems. To do that, Nexus defines a standard way for diagnostic equipment to communicate with microprocessor chips. Nexus plugs allow emulators, debuggers, logic analyzers, and workstations from one maker to control and debug processors from any other maker.*

Nexus adds a number of innovative features in one industry standard.[g] What is most interesting to me are these three unique features of the Nexus standard:

[f]https://nexus5001.org/nexus-5001-forum-standard/.

[g]Please give me some latitude here for bragging rights. I was on the steering committee for the group for a number of years and was involved in the original definition of the standard.

1. JTAG interface: The forum adopted the industry-standard JTAG port as the fundamental hardware interface protocol. The connector is also referred to as the test-access port or TAP.
2. Scalability: The Nexus standard defines 4 levels of features and performance, Class 1 through Class 4. Class 1 support is the minimum feature set that you might expect to see in any simple debugger. It requires the least amount of resources and connector pins. Class 4 is the most feature-rich and requires the addition of extra pins to the TAP to enable high-speed trace capabilities. Class 2 and Class 3 fall in between the 2 extremes.

   Within each class, there are features that must be included and others that are up to the vendor and the customer. This leads to the last unique feature.
3. Private messaging: The Nexus standard has a clever solution for the problem that customers are in competition with other customers and vendors are competing with other vendors. So, how do you have one standard that still fosters competition among the adopters of the standard?

Imagine what would happen if Vendor A implemented a feature for Customer B that was critical for the customer's competitive position against Customer C. Because the Nexus tool is a standard, Customer C could take advantage of the capability that Customer B paid for. This is where private messaging comes in.

The special feature is implemented in such a way that unless the message was understood by the tool or the CPU, it was ignored. Thus, a private message would not upset the link between the tool or the CPU; it would simply be ignored unless the tool or CPU knew about it. Thus, competitive features could be built into Nexus-compliant devices yet robust competition could still exist between vendors and between their customers.

Here's an example. Although I can't go into particulars, one vendor-customer private feature enabled the customer to tune the vibration out of their car's drivetrain while the car was running.

Here's a table that summarizes the basic Nexus features by Class [5]:

Feature	Class 1	Class 2	Class 3	Class 4
Read and write registers while in debug mode	×	×	×	×
Read and write memory while in debug mode	×	×	×	×
Enter debug mode from reset	×	×	×	×

Feature	Class 1	Class 2	Class 3	Class 4
Enter debug mode from user mode	×	×	×	×
Exit debug mode to user mode	×	×	×	×
Single-step instruction; reenter debug mode	×	×	×	×
Stop on breakpoint; enter debug mode	×	×	×	×
Set breakpoint or watchpoint	×	×	×	×
Device identification	×	×	×	×
Notify of watchpoint match	×	×	×	×
Monitor process ownership in real time (ownership trace)		×	×	×
Monitor program flow in real time (program trace)		×	×	×
Monitor data writes in real time (data trace)		×	×	×
Monitor data reads in real time			Optional	Optional
Read and write memory in real time			×	×
Execute program through Nexus port (memory substitution)				×
Begin trace on watchpoint				×
Begin memory substitution on watchpoint				Optional
Low-speed I/O port replacement		Optional	Optional	Optional
High-speed I/O port sharing		Optional	Optional	Optional
Transmit data acquisition			Optional	Optional

For the most part, these features are easily recognizable as parts of any embedded processor debugger. One feature might not be so obvious, "port replacement." Port replacement is a Class 2 and higher feature that requires that the debugger be able to replicate the behavior of up to 16 general-purpose I/O ports. This feature is a concession to the reality that the Nexus interface is eating up I/O pins that might otherwise be used for I/O purposes.

The industry committee that supports the Nexus standard, IEEE-ISTO 5001, ceased active involvement in the standard several years ago, and the standard has since been frozen with the release of version 3.0 in 2012. The 3.0 version added support for the Xilinx Aurora[h] high-speed serial protocol in order to support real-time trace information from multiple on-chip function blocks or multiple processors. According to the Nexus standard [5]:

Aurora is an industry standard (open) light-weight link protocol - ideal for a high-speed serial debug link. Nexus adheres to the Xilinx Aurora Protocol Specification V2.x.

Note that the Nexus website provides a link to the standard that was supposed to be available to members only, but I was able to download it without entering my Nexus membership credentials.

Before we wrap-up this chapter, let's examine a few more examples of on-chip debug resources.

MIPS EJTAG

The full MIPS EJTAG specification [6] is a 220-page PDF document that covers the basic and extended features of the MIPS implementation of the on-chip debug interface. The MIPS implementation of EJTAG begins with the basic infrastructure of the IEEE 1149.1 JTAG standard.

According to the MIPS specification,

On-Chip Debug (OCD) provides a solution for all these issues, and the EJTAG Debug Solution defines an advanced and scalable feature-set for OCD that allows debugging while executing CPU code at full speed.

OCD puts the ICE functionality on the chip. Although OCD does add a little extra die area for features that are only required during development, the die area is minimal. More importantly, with development time and overall time-to-market becoming increasingly critical, the trade-off between die area and time seems reasonable.

It is interesting to note here how the EJTAG specification takes aim at the functionality once only possible to implement using ICE-based tools.

[h] https://www.xilinx.com/products/intellectual-property/aurora64b66b.html.

EJTAG is an extended JTAG, which takes the basic JTAG functionality of IEEE 1149.1 and adds extensions to the standard while maintaining backward compatibility. One particularly interesting feature of the MIPS implementation is the fast debug channel (FDC) that acts very much like DMA data channels that are commonly used with disk drives. With the fast debug channel, the users set up a data transfer in much the same way as a DMA data transfer is established. Once the transfer is set, the CPU resumes normal operation and the transfer takes place through the JTAG port as a background operation.

The mechanism is implemented using a first in, first out (FIFO) memory controller. The FIFO block is memory-mapped into the CPU's physical address space. When data need to be transferred out to the JTAG port, the CPU writes the data to be sent to the FIFO transit block and then goes back to normal operation. The reading out of the data takes place in the background until the buffer is empty. The primary advantage of this mechanism is that the CPU does not have to block operation while a debugger data transfer is taking place.

Data being sent to the CPU can generate a fast debug interrupt to cause the CPU to read in the data, or the CPU can periodically poll a status bit to see if there is data to be read.

MIPS processors implement program tracing through a specification [6] that a user may implement as an additional on-chip functional block as part of a system-on-chip design. Thus, it is up to the user to create the trace circuitry and provide the interface linkage to the CPU innards as provided in the specification. This will be discussed in the next chapter.

Another advantage of on-chip debug is the ability to debug complex integrated circuits containing multiple CPU cores. With each CPU core having its own internal debug block, it would be possible to gain visibility into the system in a way that was previously impossible. We'll discuss this in the next chapter as well.

Final remarks

The great revolution in debugging embedded microprocessors occurred when the semiconductor companies decided to take tighter control of their tool chain by severing their dependency on external tool vendors to provide the sophisticated debugging tools that are unique to real-time systems. The primary casualty in this move was the companies building in-circuit emulators.

No longer would extremely talented engineers be needed to figure out ways to build an integrated instrument that simultaneously could provide classical debug features, real-time trace, and overlay memory in one package, and do it in a way that made its operation transparent to the end user.

By adding on-chip debug capabilities to every chip that leaves the semiconductor factory, the engineering problem becomes an order of magnitude easier to solve. Using a standard and open interface specification, any probe could tap into the debug core of the CPU and debug it. Adding trace capability meant that the functionality of a logic analyzer, whether internal such as in an ICE or external such as in a standalone logic analyzer, could bring the built-in capability of the on-chip debug to an even higher level of usability.

This is not to say that the in-circuit emulator is an obsolete technology. Companies such as Ashling and Lauterbach have taken advantage of the capabilities of the Nexus standard and provide advanced debugging capability, including program trace, within their tool suites.

Referring to Fig. 7.3, the Vitra-PPC is an emulator that utilizes the Nexus 5001 on-chip debug interface to provide the classic run control and trace capabilities of the classical ICE.

The emulator can support a wide range of PowerPC processors because all the family contain the Nexus debug core, so they all can connect to the single ICE unit. This is clearly an advantage

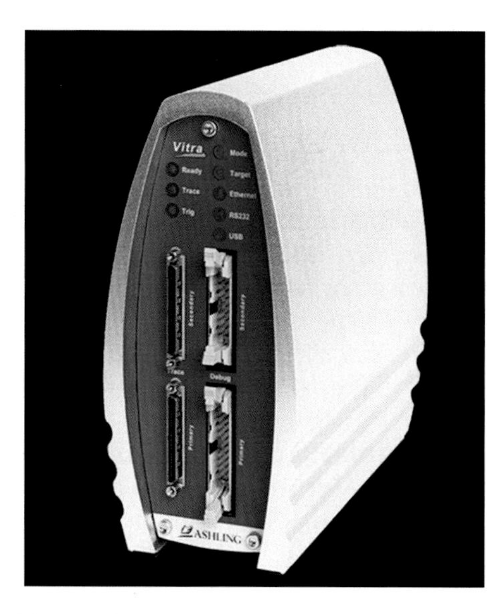

Fig. 7.3 Ashling Vitra emulator for the PowerPC family. Courtesy of Ashling Microsystems.

over the traditional ICE unit that, at best, could only support a limited number of processors (typically one) without modification.

The only capability of a full-featured emulator that is not implemented in these types of emulators is overlay memory. Theoretically, it would be possible to build into a debug core the ability to trap memory reads and writes destined to internal memory, of off-chip memory, and send those to emulation memory within the emulation system. Of course, it would likely not be close to real time, but it could be implemented.

This would have the advantage that it would work with internal RAM and ROM, something that can't be done with on-chip memory using traditional ICE tools. So, if it is technically possible to implement overlay memory, why hasn't it been done? My guess is that market research has determined that it just isn't necessary. Traditional embedded system architectures featuring RAM and ROM that are off-chip on the PC board are not as frequent as they once were, and tracing ROM can be accomplished using a ROM emulator.

Back in Chapter 5, I referenced an article that Larry Ritter and I wrote in 1995 called "Distributed Emulation." This was during my tenure at AMD when I was responsible for development tools support for AMD processors, namely the AM29000 family and the AM186 family. Emulation support for the AM29000 was weak. The article was intended to show potential users of the 29 K that the functionality of the ICE could be circumvented by a combination of other development tools.

The 29 K had a rather clever debugging function called "traceable cache." Two identical 29 K processors were mounted on a PC board (called a logic analyzer preprocessor module) with their status and data buses tied together. Using the JTAG port, one of the processors was put into slave mode. The two processors ran in lockstep with the master processor, fetching from memory and the address bus of the slave processor connected to the logic analyzer.

As the master processor fetched code from memory, the slave processor would output the current value of its program counter. With the program counter information and the visibility of the data bus, the activity in the cache could be inferred by postprocessing the logic analyzer trace data.

Certain types of code faults cannot be found without some variant of a trace capability. My favorite example is the global variable that keeps getting clobbered. It isn't enough to set a breakpoint on the variable's address because the fault only occurs 1 time in 10,000. A trace buffer that selectively stores only a few memory cycles just prior to the variable write will often point to the errant code or errant O/S task that is creating the problem.

In summary, the key issue for debugging embedded microprocessors, microcontrollers, or embedded cores always comes down to visibility during real-time code execution. As discussed, without the insight into the interplay of the operating code and the external events as these events occur in real time, debugging a heavily loaded system may be impossible.

Once we can see into the potential source of a bug, then we need to employ the standard investigative techniques that have been discussed in earlier chapters. You know the drill: "divide and conquer," etc.

In my opinion, or IMHO,[i] and with great deference to Sherlock Holmes, debugging really comes down to some very fundamental practices, insight, and experience. In terms of fundamental practices, I would list:

1. Understand how to use the tools you have available.
2. Be anal about keeping notes on your insights, observations, tests, and results.
3. Change one thing at a time, observe the behavior, makes some notes, move on.

Today, we have another dimension of powerful debugging capability that is a relatively recent phenomenon. I'm sure everyone reading this has used it. It's called the Internet. When your personal computer misbehaves after the latest Windows update, what's the first thing we do? Right, we use our favorite search engine to see if anyone else has seen the problem, and what they did to solve it.

I belong to several online interest groups and when I'm stumped, I post a question and magically, somewhere in the world someone responds to my question with the answer or an insight, or at least some encouragement.

I can't make an off-hand comment in class without a student looking it up. I can't come up with a homework problem that my students can't find the answer to online. I often wish that this resource didn't exist. I fear that the pleasure of losing track of time while grinding through a circuit problem and then having the answer emerge from a sheet of calculations is being lost on a generation. Perhaps the ability to think deeply and critically is no longer necessary in a world where we only need to find the right URL. Anyway, flame off. On to the next chapter.

[i]I am a big fan of the standard list of text acronyms.

References

[1] Computer History Museum, Intel 8051 Microprocessor Oral History Panel, CHM Reference number: X5007.2009, http://archive.computerhistory.org/resources/text/Oral_History/Intel_8051/102658339.05.01.acc.pdf, 2009.

[2] A. Berger, M. Barr, Introduction to On-Chip Debugging, Embedded System Programmingvol. 16, (2003) No 3, https://www.embedded.com/electronics-blogs/beginner-s-corner/4024528/Introduction-to-On-Chip-Debug.

[3] Texas Instruments Semiconductor Group, IEEE Std 1149.1 (JTAG) Testability Primer, SSYA002C, http://www.ti.com/lit/an/ssya002c/ssya002c.pdf, 1997.

[4] Nexus, Nexus Standard Brings Order to Microprocessor Debugging, A White Paper From the Nexus 5001 Forum, IEEE-ISTO, http://nexus5001.org/wp-content/uploads/2015/01/nexus-wp-200408.pdf, 2004.

[5] IEEE-ISTO, The Nexus 5001 Forum™ Standard for a Global Embedded Processor Debug Interface Version 3.0, IEEE-Industry Standards and Technology Organization (IEEE-ISTO), Piscataway, NJ, 2012, p. 92.

[6] MIPS Technologies, MIPS® EJTAG Specification, Document Number: MD00047, Revision 6.10, http://www.t-es-t.hu/download/mips/md00047f.pdf, 2013.

8

Systems on a chip

Introduction

When we refer to a system on a chip (SoC) or a system on silicon (SoS), we are simply acknowledging that when you have the ability to place one or more CPU cores on a silicon die and also place RAM, nonvolatile memory, peripheral devices, etc., on that same die, you've built a system. If that silicon die happens to be a field-programmable gate array (FPGA) with an embedded CPU, such as an ARM, core, then you have a reconfigurable system on a chip.

Debugging such a system provides some unique challenges, primarily the lack of internal visibility. This means that we have to depend more upon simulation tools as well as an on-chip debug core to substitute for our inability to connect a logic analyzer to the internal buses.

Another challenge is that designers can put multiple cores on one die and come up with devilishly clever ways to interconnect them to share the workload for maximum efficiency. Unfortunately, the tighter the coupling between the processors, the more difficult it will be to sort them out.

Quite a while ago, I attended one of the annual Microprocessor Forums in Silicon Valley. I recall that one presenter[a] related that his company can design an ASIC with 64 32-bit RISC processors on it and they have no idea how to debug it.

Fast forward. I just watched a YouTube video of the Tilera GX-72 SoC with 72 64-bit cores [1]. In disbelief, I went to the company's website and found this introductory blurb,

> *The TILE-Gx72™ Processor is optimized for intelligent networking, multimedia, and cloud applications, and delivers remarkable computing and I/O with complete "system-on-a-chip" features.*

[a] I truly wish I could cite a reference, but I searched and couldn't find one. You'll have to trust me that I actually heard the speaker say this.

Debugging Embedded and Real-Time Systems. https://doi.org/10.1016/B978-0-12-817811-9.00008-9

> *The device includes 72 identical processor cores (tiles) interconnected with the iMesh™ on-chip network. Each tile consists of a full-featured, 64-bit processor core as well as L1 and L2 cache and a non-blocking Terabit/sec switch that connects the tiles to the mesh and provides full cache coherence among all the cores. [2]*

Wow! I'd love to see what one of those chips could do running the thermostat in my home.

I was a member of the R&D team at Hewlett-Packard's Logic Systems Division in Colorado Springs, Colorado, that developed HP's 64700 family of in-circuit emulators. This emulator family was a departure from the standalone workstation design of the original HP 64000 because the host computer was a personal computer, linked to the emulator over an RS-232 or RS-422 serial connection.

We received a patent[b] for a feature we added called the CMB, or the Coordinated Measurement Bus. The CMB was added because we could see that debugging systems containing multiple microprocessors would become an issue that would need to be addressed by the embedded tool vendors as the technology evolved.

The CMB enabled multiple emulators to cross-trigger their internal trace analyzers as well as start together and cross-trigger breakpoints. I thought it was a pretty neat feature until we actually tried to use it. The problem was not the technology, which really worked well. We discovered that after we had more than two emulators linked together, trying to wrap our minds around what was happening became exponentially more difficult. I suspect that our customers had a similar problem.

I mention this because in this chapter we'll look at debugging integrated circuits containing multiple CPU cores, and I just want to give you an advanced warning to be sensitive to how these tools deal with trying to come to an understanding of how the cores are interacting.

Field-programmable gate arrays

For many years, FPGA was a solution in search of a problem. I don't believe this is still the case. The FPGA was originally designed to be the prototyping tool for engineers designing ASICs, and I'm sure that there is still a significant number of FPGAs used for that purpose.

[b]US patent #5,051,888.

A network of FPGAs could, in theory, be used to simulate any digital system, no matter how complex, assuming that you could interconnect them and program them, which is not an easy task. My last project at HP before I left the company was to work on a hardware accelerated simulation engine based upon a network of 1700 or so custom-designed FPGA circuits, called PLASMA chips.[c] Unique to the PLASMA chip was a large interconnection matrix and a way to program multiple chips in a large array, thus addressing the two main issues of building a big array. The machine was called Teramac [3], and it was one of the most fun projects I ever worked on.

Triscend Corporation was a pioneer in reconfigurable hardware with embedded microcontroller cores. They were purchased by Xilinx in 2004 [4], marking the point where embedded cores surrounded by an FPGA sea of gates entered the market. Today, both Altera (now owned by Intel) and Xilinx both offer FPGAs with embedded cores, primary those from ARM.

The Xilinx Zynq UltrScale+ EG features four 64-bit ARM cores running up to 1.5GHz as well as an included GPU and all the reprogrammable logic you could ask for and a full suite of development tools. At the opposite end, they offer 8- and 32-bit soft cores that can be added to any Xilinx FPGA. The MicroBlaze 32-bit RISC softcore is a good example of this. A block diagram of the MicroBlaze CPU soft core is shown in Fig. 8.1.

Not shown in the block diagram is the included debug core that provides most of the functionality discussed in Chapter 7. For smaller FPGAs, Xilinx offers an 8-bit PicoBlaze softcore that will fit into most Xilinx FPGAs. I don't want to seem like I own stock in Xilinx here, it just happened to be the first FPGA company I researched for this chapter.

Altera, the other major FPGA vendor, was acquired by Intel in 2015. The Intel Agilex and Stratix families of SoC FPGAs are similar to the Xilinx parts I discussed in having quad 64-bit ARM cores. Even at the low end, the Cyclone V SoC FPGAs have a dual-core ARM Cortex-A9 processor as well as other embedded peripherals and memory in a hard-core within the FPGA. We use this particular FPGA to teach our introductory digital electronics class on the Terasic DE-1 development board[d] and are currently designing our microprocessor class to take advantage of the embedded ARM core.

Using FPGAs as an SoC provides some very unique possibilities to address the issue of debug visibility. Many cores, whether soft or

[c] Programmable logic and switch matrix.

[d] https://www.terasic.com.tw/cgi-bin/page/archive.pl?No=83.

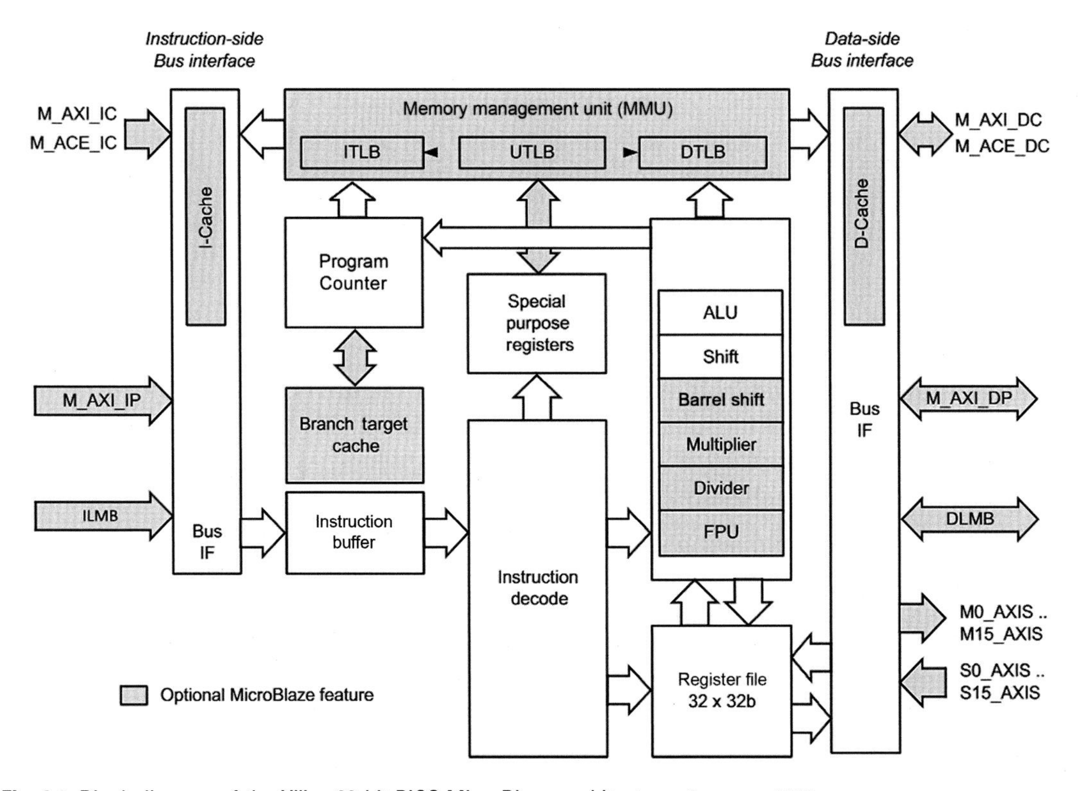

Fig. 8.1 Block diagram of the Xilinx 32-bit RISC MicroBlaze architecture. Courtesy of Xilinx.

hard, have debug blocks that are already there, or can be compiled into the design as needed. As we've previously discussed, a debug block gives the on-chip equivalent of a software debugger. If you need trace, it would not be that difficult to create a trace block with some simple address or data matching trigger circuits that would send data out to I/O pins that could be connected to a logic analyzer.

Of course, software simulation tools, such as the free Intel Quartus Prime Lite Edition,[e] come with their Signal Tap logic analyzer as a standard feature. This software simulation provides the functionality of a real hardware logic analyzer. However, when combining software running on the core and the peripheral devices, the software-only solution will likely be too slow, or completely unable to provide the logic analyzer view that a software developer would need.

[e]https://www.intel.com/content/www/us/en/software/programmable/quartus-prime/download.html.

Getting back to using an external logic analyzer, the issue would likely come down to the number of I/O pins that are available to bring out the bus signals, assuming that the bus is visible. The MicroBlaze processor in Fig. 8.1 has both I-Caches and D-Caches on-chip, so these caches would have to be disabled in order to force the core to fetch instructions and data transfers from external memory. This would have the effect of running the processor core in real time, but at a lower performance level than with the caches enabled.

In terms of a process, trying to debug an SoC in an FPGA is rather similar to debugging any real-time system. You observe the symptoms (it doesn't work, or it works but not well, or it seems to work but the results are incorrect) and form a hypothesis about what might be causing the problem. What is different about the SoC is that you are likely to have a much, much greater need for using simulations than a board-level system might require.

We are all familiar with software that has real-time timing constraints. Writing an algorithm in a high-level language such as C++ means that you are placing your trust in the compiler to create correct code. However, we realize that compiler overhead means that time-critical functions may not run as efficiently as possible, so these critical modules are often hand-crafted in assembly language.

When we expect an SoC, or a hardware algorithm, to be able to run at a certain minimum clock frequency, we will once again be putting our faith in the ability of the routing software to map our Verilog-based design into the FPGA as efficiently as possible. Sometimes, this will not be good enough, and even though the design can be fitted into the FPGA, the ultimate clock speed will be determined by the longest path through the combinatorial logic. This could include the propagation delays as well as the path length delays as the critical signal moves through the device.

As the utilization within the FPGA goes up, the number of available paths will diminish. The result is that some paths will be more roundabout due to the lack of direct paths between the logic blocks used in your design. However, unlike hand-crafting assembly code, hand-routing critical paths in an FPGA will likely not be possible, or not recommended.

We can illustrate this problem with a simple diagram. Fig. 8.2 shows a simplified schematic diagram of a pipeline, although it could just as easily be a simple finite state machine. The principle is the same. For simplicity, we'll fold any path length delays into the propagation delays within the combinatorial logic. Suppose that the set-up time for the flip-flop is 500 ps, the propagation delay through the flip-flop is 1 ns, and the slowest combinatorial block in the pipeline has a propagation delay of 4 ns, then the total delay per stage would be $1+4+0.5=5.5$ ns. Taking the reciprocal

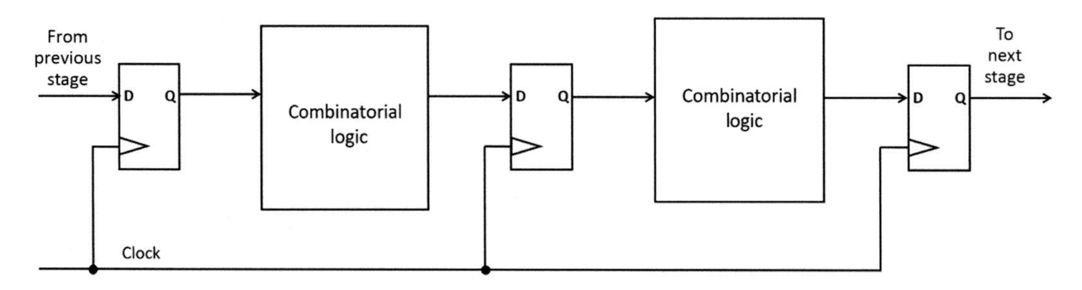

Fig. 8.2 Schematic representation of a pipeline. Maximum clock speed is determined by the sum of the propagation delay through the D-type flip-flops, the combinatorial logic, and the set-up requirement for the D input to the flip-flop.

of 5.5 ns, we find that the maximum allowable clock frequency is 182 MHz.

A good example of this discussion [5] is shown in the following Verilog code block:

```
module timing (
    input clk,
    input [7:0] a,
    input [7:0] b,
    output [31:0] c
);

reg [7:0] a_d, a_q, b_d, b_q;
reg [31:0] c_d, c_q;

assign c = c_q;

always @(*) begin
  a_d = a;
  b_d = b;

  c_d = (a_q * a_q) * (a_q * a_q) * (b_q * b_q) * (b_q *
b_q);
end
always @(posedge clk) begin
  a_q <= a_d;
  b_q <= b_d;
  c_q <= c_d;
end

endmodule
```

When this module is placed into a Xilinx FPGA, the ISE software successfully routed it but failed the timing constraint of a 20 ns maximum data path delay with a calculated delay of 25.2 ns. According to Rajewski, the reason is in the bold line:

```
c_d = (a_q * a_q) * (a_q * a_q) * (b_q * b_q) * (b_q * b_q);
```

This line involves many multiplication operations and these multiplications will be instantiated as a complex network of combinatorial logic that is simply too slow to meet the design requirements.

Could we fix this? Perhaps. We might reconfigure this calculation as a pipeline and split the multiplication operation into two steps. This would reduce the propagation delay but could add complexity in other ways.

Because an FPGA is a reprogrammable device, we would assume that the underlying part is inherently stable, assuming that we don't violate good digital design practices that might lead to metastable states and unpredictable behavior. Contrast this with a custom IC that needs to be validated to a much higher level of rigor.

A good friend of mine related this story to me. He was working for a famous supercomputer company. They made heavy use of FPGAs in their latest design. In order to wring the last bit of performance out of the system, the PC boards containing the FPGAs were hermetically sealed and each FPGA was continuously sprayed with high-pressure Freon to keep it cool. Even with this heavy driving, the FPGA was still being driven within its band of allowable timing specifications. The computer design team noticed that occasionally one of the FPGAs would appear to flip a configuration bit and go off into the weeds.

They consulted with the FPGA manufacturer, who was quite skeptical that any application could cause bit flipping to occur. It was only when they saw the computer in operation and observed the failure did they believe that they had a problem and needed to redesign the part to make it even more tolerant to switching transients. The moral of this story is that even production parts can suffer from glitches under the right conditions.

One interesting aspect of FPGAs that I have not seen much written about is the concept of using the reprogrammability of the hardware environment surrounding the embedded core(s) to build various kinds of hardware-based debug tools. These tools can be looked upon as the hardware equivalent of turn-on or throw-away code to a software engineer. Of course, it is unlikely that once you develop a neat hardware debugger in Verilog, you then toss it. Once you have it and it works, then there is a real-time debugger available anytime. For example, the real complexity inherent in a logic analyzer is tied to the triggering circuitry. The Hewlett-Packard logic analyzers I am familiar with had incredibly flexible (and complicated) trigger circuitry, if you want to use it to track down an elusive bug that didn't lend itself to a simple break point at an address. This could be done through sequencing, much like a finite state machine circuit.

The HP logic analyzers had seven levels in the sequence. At each level, you could do logic combinations of any number of the bits being used. You could count cycles as well. When a TRUE result occurs, the sequencer could trigger the analyzer to record data or move to the next state in the sequence. A FALSE result could cause the state to remain the same or take you back to a prior state to start again.

But suppose all that you need to do is trigger on an address, data, or status value. It would be relatively simple to build a simple breakpoint comparator in Verilog that would trigger on any combination of bits. The trigger signal could be used to generate a processor interrupt or to trigger a circular trace buffer. The circular buffer is a bit more complicated than a linear buffer, but it has the advantage of being able to capture events leading up to the trigger point as well as what happened after the trigger, or anything in between. Fig. 8.3 illustrates how the circular buffer works.

In Fig. 8.3, the buffer holds 2^{24}, or roughly 16 million, states. Each memory cell can contain the total number of bits in your logic analyzer. The LA that I am most familiar with could record up to 192 bits wide, although the memory depth was much smaller than 16M.

In our example, we would have a 24-bit binary counter generating the memory addresses. An address comparator determines where in the buffer we want the trigger point to occur. Thus, the trigger address would be in the range of 000000H to FFFFFFH.

In normal operation, the trace buffer is running continuously and once it records 16M states, it will start to overwrite previously written data. Another 24-bit countdown counter is also part of the circuit, but is not running until the trigger signal occurs. At that

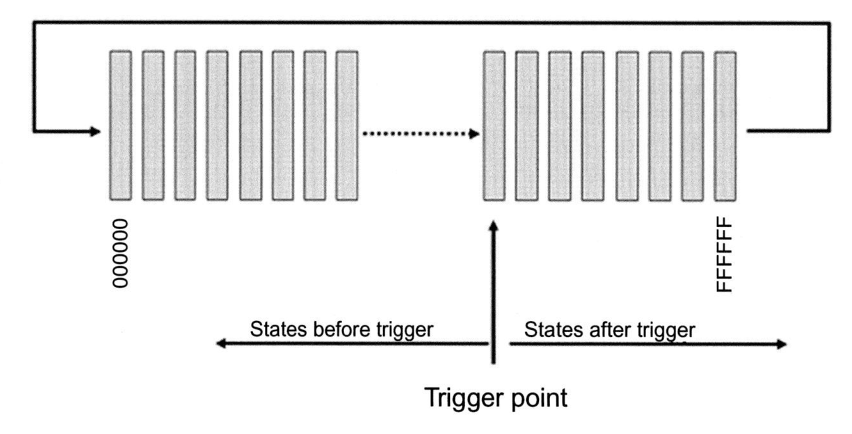

Fig. 8.3 Schematic representation of a circular trace buffer architecture used in a logic analyzer.

point, it will start to count down from 16M to 0 and when it reaches 0, it turns off the write signal to the memory buffer, effectively preventing the data from being overwritten. To read out the buffer, reset the memory address counter and read out the data.

Do you need 16M states and 192 bits in width? Maybe. But likely, you'll require a subset of the width and a simple buffer to capture the data.

The logic analyzer could be constructed as a peripheral of the embedded core and can be through your debug or validation software. Once you are done with your logic analyzer, some simple additions, perhaps a 64-bit Gray Code counter, can be added to the circuit. Then you can turn the logic analyzer into a real-time performance analyzer.

It gets even better. If you have a spare CPU core in your design, turn that CPU into your logic analyzer controller. Keep it entirely separate from the primary CPU. Here's an example of the Altera (oops, now Intel) Cyclone V FPGA that I previously mentioned. The text is from the introduction to the FPGA family of parts: [6]

> *The Cyclone V SoC FPGA HPS consists of a dual-core ARM Cortex-A9 MPCore* processor, a rich set of peripherals, and a multiport memory controller shared with logic in the FPGA, giving you the flexibility of programmable logic and the cost savings of hard intellectual property (IP) due to:*

- Single- or dual-core processor with up to 925 MHz maximum frequency.
- Hardened embedded peripherals eliminate the need to implement these functions in programmable logic, leaving more FPGA resources for application-specific custom logic and reducing power consumption.
- Hardened multiport memory controller, shared by the processor and FPGA logic, supports DDR2, DDR3, and LPDDR2 devices with integrated error correction code (ECC) support for high-reliability and safety-critical applications.

If you want to really see what is going on, you can turn the logic analyzer from a state analyzer, where the state acquisition occurs synchronously with the CPU clock, into a timing analyzer by driving it with a separate clock, ideally running faster than the CPU clock. The key idea here is that internal visibility into an SoC based upon the FPGA architecture is possible to achieve.

I've often had discussions with my colleagues regarding my insistence on making real measurements versus simulations. This usually centers around my focus on teaching how to use a logic analyzer in my microprocessor class. Someone will point to the data sheet for some FPGA simulation tool and show me the built-in logic analyzer that runs within the simulation. So, why bother with measuring real circuitry, just run simulations?

I have to admit that they have a point. Simulations are getting better and better, particularly the FPGA design tools, and the ones we use are free for the downloading! What a deal. But…. logic analysis, as a debug methodology, is so fundamental to what a hardware engineer needs in her toolkit that I just can't imagine relegating it to the pile of obsolete electronics that you find in the surplus stores.[f]

Lauterbach GMBH offers one of the most extensive set of development and debug tools specifically designed for debugging embedded cores. Their TRACE32 debug tool family supports an impressive range of hard and soft embedded cores for both trace and debug. The trace buffer can be configured on-chip or through a parallel port to a trace buffer on the host computer, which provides an almost insanely large trace buffer of 1 T frames. Fig. 8.4 is a schematic diagram of the Lauterbach system.

As you can see, multiple cores can be connected to on-chip trace generation logic, as I've described, then to the trace buffer. This information can then be downloaded through the JTAG port to the TRACE32 user interface to be analyzed in the same way that you would analyze a board-level embedded system. The flexibility of the FPGA enables logic analysis modules to be added for critical debugging stages and removed if the space is needed for additional functionality. Intel, Lattice, and Xilinx offer configurable

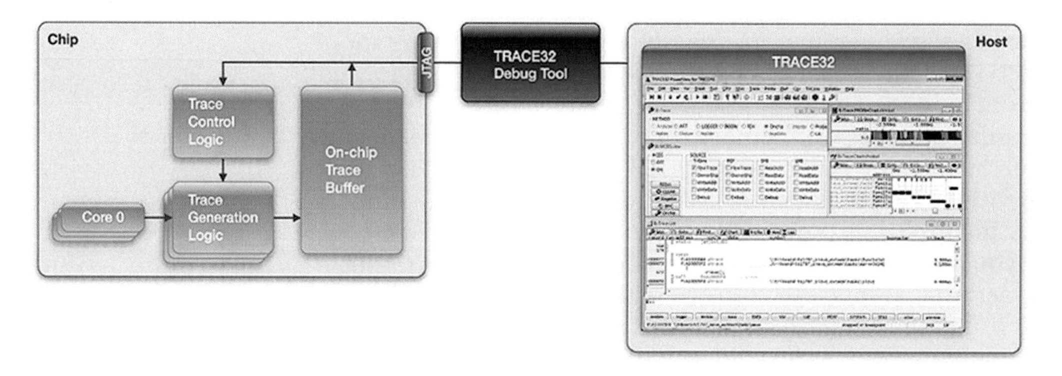

Fig. 8.4 Schematic representation of the TRACE32 debugging configuration. Courtesy of Lauterbach, GMBH.

[f] I was once in a surplus store in Silicon Valley and came upon an HP scope that I had worked on. It was available as a nonworking, parts only hulk for $20. I got so upset, I walked out.

modules for integration into their FPGA families. Intel offers the Signal Tap Logic analyzer. Lattice offers Reveal and Xilinx offers the ChipScope analysis blocks.

Because you can expect to pay for the commercial LA blocks, I was curious to see if there were any public domain logic analyzers that I could recommend. I found one such homebrew on a blog [7] and the authors reference some other work. However, be forewarned. Playing with open source software may be a bigger time suck than just purchasing a license to use the commercially available modules.

Gisselquist [8] describes in a rather complete set of instructions how he built a 16-bit in-circuit logic analyzer in Verilog. He goes step by step through the subblocks and then provides an example application.

Another excellent example [9] was a final project of Mohammed Dohadwala, a student in ECE 5760, Advanced Microcontroller Design, at the Electrical Engineering School of Cornell University.[g] This system was implemented on a Terasic DE-1 FPGA Board[h] that was based on the Intel (Altera) Cyclone V FPGA, which contains an ARM dual-core Cortex A9 hard core. The full report is available online (see reference). What I like about his design is that it uses a 32-bit wide, 512 element deep FIFO (memory), rather than a standard trace buffer. This enables him to run the logic analyzer at 100MHz while still being able to stream the data to an external data logger. A key element of the design is an IP block called XILLYBUS,[i] which provides a DMA function over PCIe I/O protocols. The bus is designed to work with Xilinx Zynq-7000 EPP and Intel Cyclone V SoC.

This logic analyzer is designed to be controlled by the onboard ARM core and runs under Linux. As such, its primary use would be to debug the other functional blocks in the system rather than the CPU core. Fig. 8.5 is a block diagram of the system.

Virtualization

Today, we have PCs that have performance numbers that the workstations of 10years ago could only dream about. With that computing power, multiple gigabytes of RAM, and fast solid-state drives, it is possible (with the appropriate software) to completely

[g]My alma mater.

[h]This is the same FPGA board that we use in our EE program.

[i]www.xillybus.com.

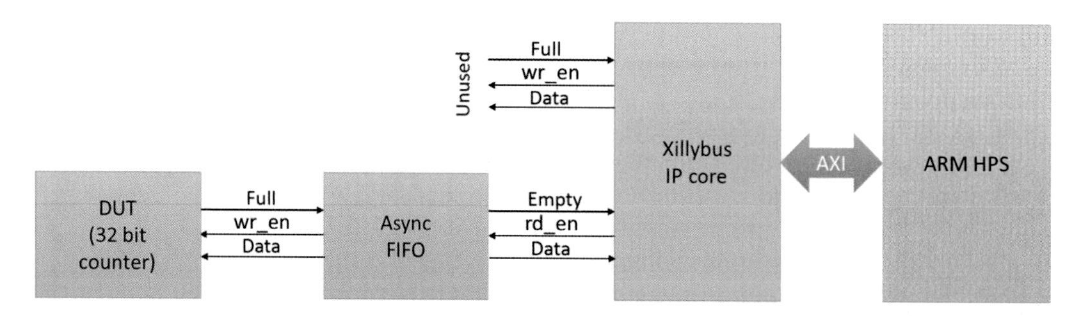

Fig. 8.5 Block diagram of a logic analyzer implemented as a final project for an EE class at Cornell University.

model a system, whether it is an SoC or a board-level system with discrete components.

Virtualization is not a new technology, per se. Instruction set simulators have been around for a long time. Apple iMac computers can run Windows software in a virtual PC. The key to the major advancement in the technology is the idea of a hypervisor layer that sits beneath the processor and its supporting hardware. This hardware includes memory, so in a multicore environment or a multiprocessing environment, each core or virtual processor can run independently of the others. This is convenient if, for example, you wish to run different operating systems on each core. Another virtue is that it provides a level of security that prevents a hacker who may gain access to one virtual machine to then get a jumping off point to other virtual machines running high-security applications.

In a Technology White Paper, Heiser [10] describes how his OKL4 kernel[j] hypervisor can protect a communications stack in a cell phone from a virus-infected application. He goes on to point out that even with multiple cores, there is a danger if the cores share a common memory bank.

Another name for the hypervisor is the virtual-machine monitor, or VMM. According to Popek [11] and cited by Heiser, the VMM must have three key characteristics:

1. The VMM provides to software an environment that is essentially identical with the original machine.
2. Programs run in this environment show, at worst, minor decreases in speed.
3. The VMM is in complete control of system resources.

[j]The OKL4 kernel is now offered as a commercial product by General Dynamics Mission Systems. See https://gdmissionsystems.com/en/products/secure-mobile/hypervisor.

Condition 1 guarantees that the software designed to run on the bare hardware (actual machine) will also run (unchanged) on the virtual machine. Condition 2 is important because if the virtualization is to be able to run in real time, it must not suffer a performance hit so bad as to prevent time-critical code from executing properly. The third condition must guarantee that there is no back door around the VMM and that one application is completely isolated from any other.

In the OKL4 kernel running with an ARM core, the memory management unit (MMU) on the ARM core is used to provide a hardware-based isolation mechanism. The author draws a distinction between the VMM as defined by the hypervisor as a system-level virtual machine in order to distinguish it from a process-level VM, such as the JAVA virtual machine.

In embedded systems, the hypervisor enables multiple operating systems to exist concurrently. Why is this an advantage? The cell phone, for example, contains devices with real-time requirements and applications that resemble those on your PC. Thus, an RTOS can be used for the time-sensitive applications and a traditional O/S, such as Linux, can be used to control the time-insensitive applications.

So far, we've discussed virtualization as a strategy for use on a system in operation, not under development. However, it doesn't take a great leap to see that this can also be used as a powerful design environment as well. Wind River Systems[k] offers Simics, a virtual design environment that enables complete system simulation. As stated in Wind River's product overview document [12]:

> *Software developers use Simics to simulate nearly anything, from a single chip all the way up to complete systems and networks of any size or complexity. A Simics simulation of a target system can run unmodified target software (the same boot loader, BIOS, firmware, operating system, board support package (BSP), middleware, and applications as the hardware), which means users can reap the benefits of using a pure software tool.*

Fig. 8.6 shows a schematic diagram of the Simics system.

The Simics environment enables a complete system simulation that enables the software team to develop and debug their code in a continuous manner, rather than waiting until target hardware is available at the start of the HW/SW integration phase. According to the product overview document, the value of Simics

[k]www.windriver.com.

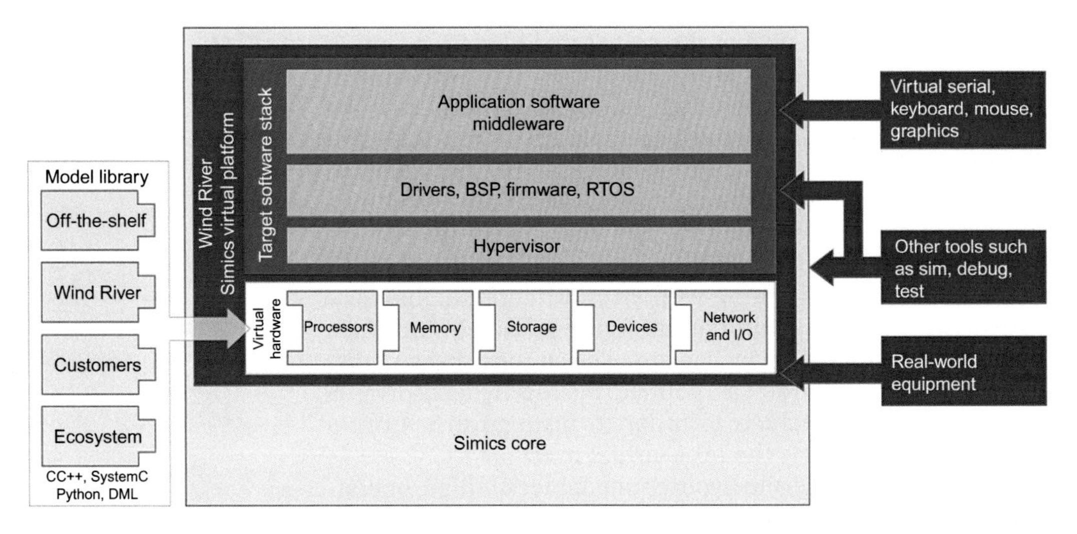

Fig. 8.6 Block diagram of the Simics simulation environment. Courtesy of Wind River Systems.

for the integration and test phase of a product development life cycle may be summarized as follows (their bullet points):

- Start testing and automation early in the development process. Do continuous hardware and software integration early, on virtual hardware, expanding to physical hardware as it becomes available.
- Build more levels of intermediate setups than are available with hardware, to facilitate continuous integration.
- Test fault tolerance with Simics fault injection. Cover corner cases that cannot be reached in hardware.
- Automate and parallelize testing and expand coverage of target configurations using Simics scripting.
- Save developer time, reduce waiting time to run tests, and shorten feedback loops by using simulation labs in addition to hardware labs.
- Do test and integration on the entire system by integrating Simics models of computer hardware with external models of the physical world or system environment.
- Automate regression testing and continuous integration by tying Simics into existing workflows of software build and test.

Of particular interest to me in the context of this book is the assertion regarding debugging:

> *Complex and connected systems are difficult to debug and manage. While traditional development tools can help you track down bugs*

related to a single board or software process, finding a bug in a system of many boards and processor cores is a daunting task. For example, if you stop one process or thread with a traditional debugger, other parts in the system will continue to execute, making it impossible to get a globally coherent view of the target system state.

Simics provides access to, visibility into, and control over all boards and processor cores in the system. Single-stepping forward and in reverse applies to the system as a whole; the whole system can be inspected and debugged as a unit. Furthermore, a checkpoint—or snapshot—can be created, capturing the entire system state. This state can be passed to another developer, who can then inspect the precise hardware and software state, replay recorded executions, and continue execution as if it never stopped.

Of course, the devil is in the details. I have not investigated the investment required in acquisition costs, licenses, training, and deployment to know if this is a good investment for any particular application, such as yours. My interest is only to help make you aware of the tools that are currently out there waiting for you. My best advice is to contact the vendors I've referenced and invite them in for a demonstration.

Another great way to see these products in action is to attend the next Embedded Systems Conference.[1] It is pretty easy to score free tickets to the exhibition floor, though it will generally cost you to attend the technical sessions. Speaking of the technical sessions, I am the proud owner of several Embedded System Conference Speaker polo shirts, which I proudly wear as my fashion statement.

If you can't, or don't, wish to attend the technical sessions, you can usually purchase a DVD of the proceedings. However, the real objective of the conference is the exhibition floor. There you can see Simics or other tools in operation and speak with engineers who are incredibly knowledgeable about the product. If you are really fortunate, you can meet with one of the R&D engineers who actually designed it, and because they do not have the brain filter of marketing and sales folks, you'll get the straight skinny, engineer to engineer, until the marketeer comes by and shoos them away (just kidding).

[1]See https://www.embeddedadvisor.com/conference/ for a listing of conferences that you might like to attend.

Conclusion

In this chapter, we've examined the tools that are relevant to trying to debug embedded cores. The issue is how do you look inside the FPGA to find and fix the bugs? Fortunately, the FPGA is an incredibly flexible device that, in my opinion, will revolutionize computing as we know it.

As an aside, during this past year, there was a lot of press coverage of the technology issues surrounding the security aspects of the Huawei data switch. While I don't know enough to know if these are real concerns, I noted in one of the technical discussions that the data switch used FPGAs as part of the machine's architecture. Thus, while it would be possible for people worried about the security of the switch to examine the CPU code, it is entirely possible for a bad actor to reprogram the FPGAs to sniff the data going through the switch. Just a thought.

The tools that we've discussed here provide the internal visibility that is needed. Once the visibility is achieved, the same debug techniques that were discussed in earlier chapters come into play. For example, keep a record of your observations, hypotheses, tests, and results. Change only one thing at a time and log any differences you might observe.

Another observation I could make is that from my examination of the tools, or at least the printed literature, I get the impression that there is likely a significant learning curve associated with becoming proficient and comfortable with them. This will require a time investment and up-front planning to allow time in your schedule for training. It can be really frustrating (I know, I've been there) to know exactly what you want to test, or do, but you can't decipher how to do it, and the documentation is of little or no help.

Onward.

Additional resources

1. https://www.eetimes.com/author.asp?section_id=36&doc_id=1284571#.
2. https://www.electronicproducts.com/Digital_ICs/Standard_and_Programmable_Logic/Debugging_hybrid_FPGA_logic_processor_designs.aspx.
3. https://www.embedded.com/design/other/4218187/Software-Debug-Options-on-ASIC-Cores.
4. https://www.edn.com/design/systems-design/4312670/Debugging-FPGA-designs-may-be-harder-than-you-expect.

5. https://www.newark.com/pdfs/techarticles/tektronix/XylinxAndAlteraFPGA_AppNote_MSO4000.pdf.
6. https://www.xilinx.com/video/hardware/logic-debug-in-vivado.html.
7. https://www.dinigroup.com/files/DINI_DR_WhitePaper_031115.pdf.

References

[1] https://www.youtube.com/watch?v=6FXMx7kvOvY.
[2] https://www.mellanox.com/page/products_dyn?product_family=238&mtag=tile_gx72&ssn=dcoil9vjj80rjjlj3p6h5ifgh4.
[3] G. Snider, K. Philip, W. Bruce Culbertson, R.J. Carter, A.S. Berger, R. Amerson, The Teramac configurable compute engine, in: W. Moore, W. Luk (Eds.), Proceedings of the 5th International Workshop on Field-Programmmable Logic and Applications, Oxford, UK, September, Pg. 44, 1995.
[4] https://www.design-reuse.com/news/7327/xilinx-acquisition-triscend.html.
[5] Justin Rajewski, https://alchitry.com/blogs/tutorials/fpga-timing, January 11, 2018.
[6] https://www.intel.com/content/www/us/en/products/programmable/soc/cyclone-v.html.
[7] Al Williams, https://hackaday.com/2018/10/12/logic-analyzers-for-fpgas-a-verilog-odyssey/, September 27, 2019.
[8] D. Gisselquist, https://zipcpu.com/blog/2017/06/08/simple-scope.html, 2017.
[9] M. Dohadwala, https://people.ece.cornell.edu/land/courses/ece5760/FinalProjects/s2017/md874/md874/LogicAnalyzer.htm, 2017.
[10] G. Heiser, Vitualization for Embedded Systems, Document Number: OK 40036:2007, 2007.
[11] G.J. Popek, R.P. Goldberg, Formal requirements for virtualizable third generation architectures, Commun. ACM 17 (7) (1974) 413–421.
[12] https://www.windriver.com/products/product-overviews/Wind-River-Simics_Product-Overview/.

9

Testing methods for isolating defects

Introduction

Just type "fault isolation" or "fault detection" into your favorite search engine and you'll get plenty of hits. Once again, many scholarly articles and books have been written about the subject, so this chapter can only skim the surface. Fault isolation is all about trying to [1]:

isolate the component, device or software module causing the error.

You could argue that everything I've discussed in Chapters 1–8 has been a set-up for this chapter. Isn't isolating the bug the key to fixing it? Therefore, let's look at various types of defects, perhaps some we've previously discussed, and see if there are specific techniques that we can draw upon to isolate them.

Of course, the problem with defects in embedded systems is that we will typically have more variables that could be the culprit than other technologies, but I'm sure that any other specialty will argue that finding problems in their systems is just as, or more challenging, than a system with untested hardware and untested software.

As we've discussed in the chapter on HW/SW integration, bringing new hardware together with new software is a major challenge, and we've looked at best practices to mitigate the risks inherent in this process.

Barriers to finding problems

We've discussed the need to have a disciplined approach to finding and fixing bugs and I've mentioned how shotgunning can be the start of your journey down the rabbit hole, never to be heard from again.

Debugging Embedded and Real-Time Systems. https://doi.org/10.1016/B978-0-12-817811-9.00009-0

I may have a Ph.D., but I'm not a psychologist (obviously) and I don't have the answer to human nature, sorry. I'm just as guilty of trying the easy stuff first, except that if my quick fix doesn't work, I'll stop that approach and bite the bullet. Out comes the scope, or the logic analyzer, the emulator, and the documentation and I go to work. And when I pull out the heavy artillery, I start keeping a log of what I'm doing and what I've observed. In my career, I've only had one instance where I haven't been able to find a bug. This is a frustrating story. I still have the errant PC board in my office and I occasionally will look at it and sigh. It just stares back, daring me to find the problem.

The board is a trivially simple Z80 board that we use to teach our EE students how to use a logic analyzer. The Z80 has one really cool feature that makes it ideal for this purpose. There is absolutely no form of performance acceleration. No caches, no prefetch queues, nothing. What you see on the bus is exactly what the processor is doing. The PCB has a 40-pin connector that plugs directly into the logic analyzer. Nothing to set up, no grabbers. We give the students a ROM with some Z80 code and ask them to figure out what the processor is doing.

Back to my problem. In order to build-up 20 or so of these boards for a class and teach the students who have never picked up a soldering iron how to solder, our local IEEE student chapter sponsors a Saturday morning pizza and soldering session. As most EEs know, pizza is the breakfast of champions and represents all the important food groups.

The students start with bare boards and solder all the parts to the board. All the parts are through-hole and easy to solder. Then we turn them on. I would say roughly 50% turn on right away. When they don't, I examine them under a magnifier, and I can almost always point to:

- Solder bridge.
- Cold solder joint.
- Missed soldering of a part.
- Parts with some legs not properly inserted, or parts backward, or bent under.

This is the usual stuff. Except for this one board. I looked for every obvious fault and everything looked good. Nice clock signal as well (I thought). I took it home where I keep the big guns and started probing with a logic analyzer. All the signals were wiggling. Set the analyzer to trigger on RESET and when the code came out of RESET the processor promptly went into the weeds.

Next, I plugged my trusty HP Z80 emulator into the board[a] and systematically tried to find the problem. Nothing worked. If I were to bill myself my hourly consulting rate, this would be the most expensive Z80 board in history. Anyway, as of this writing, 4 years later, it is still on my bookshelf mocking me. Sigh....[b]

Quick and dirty

It's human nature, or the student mentality, to do a quick fix and move on. Students are always under pressure, always behind schedule. Ditto for most engineers. I've spent my share of all-nighters in the lab at school and at my employers. You do a simple cost/benefit analysis in your head and decide that, rather than set up the tool, write test software, dig out the documentation, or whatever you think the correct approach might be, the "quick and dirty" approach is to just make an educated guess and try something. Maybe Occam's Razor will work for you this time, and Murphy's Law won't apply.

Asking for help

Sometimes, the best source of insight is the tribal knowledge that is the underpinning of your workspace and colleagues. I think that for a new engineer, there is a natural tendency to want to prove yourself to your peers. That's pretty normal. So, when you are faced with a bug that just can't be swept under the rug, you try to prove yourself by finding and fixing it yourself, no matter how long it takes. Unfortunately, as we all know, time is money and the best way to find and fix a problem is to ask the old engineer who's seen it all (like Bob Pease) to help you get some insight into the problem.

[a] I was the hardware project manager for the HP 64700 product family. We had enough scrapped parts that I was able to build myself an emulator for my home brew electronic projects, such as my hot tub controller.

[b] There is a postscript to this story. After writing this chapter, I became reinspired to fix it. This time, I started with an oscilloscope, not a logic analyzer. I unplugged every part that was socketed (a good idea when students are involved) and started with the clock oscillator. The amplitude was about 2 V! Whoa! Could it be this easy? I replaced the oscillator and the amplitude became a little under 5 V. So, I put everything back, plugged in my turn-on ROM, plugged in the logic analyzer, and the board worked fine. There are hidden benefits to writing a book.

It isn't necessary to ask for them to come to your bench and take a look. Maybe all you need to do is politely ask if you could run a problem by them. If they are the kind of engineer that believes in mentoring the newbies, then you'll be invited to sit down and discuss your problem. Now here's the good part. You whip out your lab notebook with your observations, what you did, and the results of your tests, and you share this with your mentor. Just this demonstration of professionalism will offset any negative vibes about your debugging skills.

As you discuss the problem, take notes, and most importantly, leave your phone in your desk so you won't be tempted to answer it, or respond to it, or look at it while they are talking to you. That will be the kiss of death. When a student comes to talk to me and while I'm talking, starts looking at their phone, I just ask them to leave and come back another time without their phone. I don't think it's me, although I may be stuck in a prior universe where manners meant something. Flame off.

Another good practice is to ask if you can see them again if the problem is still elusive. They'll generally say, "Sure, no problem." Although they may also tell you that they're busy and perhaps someone else has more time right now. Take that as your cue. When you find the problem, be sure to let them know and thank them. Then tell your manager how this engineer helped you. Don't take credit for someone else's knowledge.[c]

Fault isolation

Anyway, let's look at some case studies of problem areas and see if we can derive some general approaches. In order to start our analysis, we can describe some general categories. You may disagree with my selection process, but you are certainly free to use your own system for characterization.

- Performance-related faults: The system works, but not up to the required level of performance to meet your specification. The fault is not critical because it works and does what it is supposed to do, just not as well as it needs to be in order to meet its design or marketing goals. An extreme example of this is my microprocessor class where the students are given the task of designing a function generator that works up to 100 kHz and some of the designs would not work over 100 Hz.

[c] I know, I said this before, but it is worth reiterating because your circle of tech support will quickly shrink without recognition.

- Reproducible faults: The fault can be made to happen. You know how to cause it to happen, but you don't know why it is happening. Perhaps you've looked at the software and you've looked at the hardware, and everything looks normal, yet the system will fail every time. I discussed a great example of this type of fault back in Chapter 3 with the problem of a memory allocation fault in our discussion of the Y2K problem. You may recall that the root cause of the problem was the fact that by adding a four-digit date field to a data logger, the system slowly ran out of memory and there was no error handler to deal with it.
- Intermittent faults: This is the category that drives engineers to tears and causes divorces. Sometimes it happens, sometimes it doesn't. Perhaps the system will run well for an extended period of time and then crash, or else produce errant data. I would also add "glitches" to this category, but I suppose that some readers will want to place glitches in a separate category.
- Compliance faults: This is one of the most difficult categories to deal with unless you have dedicated support engineers who are expert in areas such as RF suppression or regulatory compliance. I will be the first to admit that I know very little about this field and what I do know I learned the hard way. I was also fortunate that my former HP division (Logic Systems Division, or LSD) had a very able RF compliance engineer named Bob Dockey, and I learned everything I know from conversations with him, or by watching my product fail RF testing in the anechoic chamber or during open-field testing.
- Thermal problems: This might include failure due to excessive heat, or excessive cold, or even a slight temperature deviation from ambient temperature.
- Mechanical problems: Here we have anything that has to do with the packaging and interconnections of components. Back in Chapter 4, we discussed the microphonics problem with the HP1727A oscilloscope that had me as the CRT engineer. You'll recall that when set up for a single-event capture, the scope would errantly trigger if it was jarred. In retrospect, that was likely caused by a faulty connection that opened and closed when the scope received a mechanical shock.
- Power supply issues: Noisy power and ground buses are a fruitful place to look if you are seeing glitches in your circuit behavior. This is particularly true when you have analog and digital circuits on the same board.

Notice that I did not attempt to separate software and hardware faults. That was by design because almost all these categories can be impacted by a software bug as well as a hardware bug. Don't believe me? How about thermal problems? Your faulty algorithm overdrives a transistor and it goes into thermal runaway. Well, you might argue that is a hardware design flaw because the transistor should be protected against such a fault. Maybe, and you can always make that argument during your next performance review before the scheduled 10% downsizing takes place.

The trick in all this is to use your engineering experience and insight as well as your engineering "best practices" (as we've previously discussed) to quickly drill down to find and fix the root cause problem.

Before we go back and look at some of the general categories, let's look at a few basic realities that you must deal with. There are only two general rules for fault isolation that I can make with full knowledge that they are absolutely true.

1. Know your tools.
2. Understand your design.

Know your tools

When your measurement tool touches your circuit, it becomes part of the circuit. Whether this interaction is a minor or major perturbation depends upon what you are trying to measure and what you are using to measure it. Keysight Technologies published a wonderful series of white papers, articles, seminars, webcasts, etc., on the proper use of the test and measurement products it sells. Being a former oscilloscope designer, I'm rather drawn to the o-scope as an instrument and using it properly is emphasized in our curriculum. One of my favorite articles is titled "Take the Mystery Out of Probing: 7 Common Oscilloscope Probing Pitfalls to Avoid" [2]. Here is what Keysight says in its introduction,

> *In an ideal world, all probes would be a non-intrusive wire attached to your circuit having infinite input resistance with zero capacitance and inductance. It would provide an exact replica of the signal being measured. But the reality is that probes introduce loading to the circuit. The resistive, capacitive, and inductive components on the probe can change the response of the circuit under test.*
>
> *Every circuit is different and has its own set of electrical characteristics. Therefore, every time you probe your device, you want to consider the characteristics of the probe and choose one that will have the smallest impact to the measurement. This includes everything from the connection to the oscilloscope input down*

through the cable to the very point of connection on the DUT, including any accessories or additional wiring and soldering used to connect to the test point.

According to Keysight, the seven common pitfalls are:

1. Not calibrating your probe.
2. Increasing probe loading.
3. Not fully utilizing a differential probe.
4. Selecting the wrong current probe.
5. Mishandling DC offset during ripple and noise measurements.
6. Unknown bandwidth constraints.
7. Hidden noise impacts.

In my experience, of the seven common pitfalls, numbers 1, 2, and 6 are the most common that I observed with my students. That is not to say that the other examples are less critical because each can help isolate a fault under the right conditions. For example, a differential probe can make all the measurements that you can make with the more common single-ended probe, but the differential probe can eliminate common mode noise or poor ground connections, especially at higher frequencies.

However, we do not have differential probes in our undergraduate teaching labs because students haven't yet learned how to take care of delicate instruments. In fact, upon entering our program, students must purchase lab kits that contain a general purpose 100 MHz scope probe to use with our Tektronix oscilloscopes because they tend to damage the pricy ones that came with the scopes. So, we must make do with what we have. However, a quasi-differential measurement can be made using two probes and setting the trace display for A minus B mode.

For most measurements, an uncalibrated oscilloscope probe will be fine to make a yes/no type of measurement. But when the fidelity of the waveform is important, as it was for a former student who was trying to understand why his push-pull high-voltage generator was burning out the MOSFET drivers on the transformer primary, then calibrating the probe becomes an important factor in trying to analyze the fault.

Pitfall number 2 is often an issue as well. Quoting the Keysight eBook [2],

As soon as you connect a probe to your oscilloscope and touch it to your device, the probe becomes part of your circuit. The resistive, capacitive, and inductive loading that a probe imposes on your device will affect the signal you see on your oscilloscope screen. These loading affects can change the operation of your circuit under test.

Referring to pitfall number 6, let me give an example. Our lab scopes are all 100 MHz bandwidth instruments, as are our probes. We own some 1 GHz scopes, but those are kept in a locked closet and always used under the supervision of a lab instructor. According to the Keysight article, having inadequate bandwidth will distort the signal that you are trying to measure and make it difficult to make good decisions about the fault you are trying to isolate. Keysight gives this formula for the net bandwidth of an oscilloscope/probe combination:

$$\text{System bandwidth} = \frac{1}{\sqrt{\frac{1}{\text{Scope bandwidth}^2} + \frac{1}{\text{Probe bandwidth}^2}}}$$

Let's use our student lab set-up as an example. If both the scope and the probe have a 100 MHz bandwidth, then the resulting system bandwidth is 70.7 MHz. Keep in mind that the bandwidth of an oscilloscope is defined as the point at which a sinusoidal signal is reduced by 3 dB, or about 30% of the amplitude at a lower frequency. Siglent Technologies [3] recommends that for accurate digital measurements, the oscilloscope bandwidth should be $5\times$ to $10\times$ the bandwidth of the highest fundamental frequency you are working with. This means that if your system has an 80 MHz clock, the system bandwidth should be in the range of 400–800 MHz in order to capture the higher harmonics of the pulses in the system.

In effect, the oscilloscope/probe system is a low-pass filter for the system under test. If the bandwidth is rolling off at 3 dB per octave, then at about 200 MHz, the measured signal amplitude is about 30% of what it actually is.

Now, digital oscilloscopes can muddy the waters, even though they have analog front ends that obey the above bandwidth rules. They can do this because digital scopes have all sorts of built-in calculations that they can do to adjust a measurement, particularly if the signal being measured is repetitive rather than single shot, but that's going down the rabbit hole a bit too far for now.

To summarize, finding an elusive bug or fault in your system depends upon a thorough understanding of the measurement tool being used to find the fault. Without a complete understanding of how the tool interacts with the system, or even more fundamentally, what the tool is capable of measuring, you are severely handicapping yourself and dramatically lowering the probability of finding the bug, not to mention how you are being viewed by your coworkers.

Understanding your design

Let's now consider the other major impediment to finding and isolating faults in your system. This might seem strange to you, given that you've designed it. Well, did you really design it? Are you like my students, guilty of finding a circuit on the web, or in an applications note and plopping it down into their schematic without reading the fine print about how it works and doesn't work?

What about software? Just as bad. You may want to reuse a module that's been around forever and sort of works, but the code itself is worthy of honorable mention in The International Obfuscated C Code Contest.[d] Or suppose that you are using a library function that is supposed to work. This could be C code or Verilog, same problem. It has been known to work in the past, so why not use it? Reuse is good, isn't it?

Using hardware and/or software that you did not design is an act of faith. You are trusting that it works as advertised, but if you suspect that it is faulty, where do you begin to look if you don't have the theory of operation? Of course, if you are using well-documented libraries from reputable sources, then you should be safe, unless the documentation has a typo or is confusing. Confusing Big Endian and Little Endian is a classic bug source and is particularly nasty when you are trying to access hardware by dereferencing a pointer.

So, here's a simple rule. If you discover a fault in your system under development and you are faced with a circuit, circuit element, or software module that you did not personally design, and the person who may have designed it isn't sitting in the next cubicle, then do not attempt to isolate the fault until you have studied the circuit so that you have a full and complete understanding of what you are going to try to debug.

Of course, you might do the really gross measurements, such as power supply stability and noise or ground bounce, but once they are out of the way, sit down in a quiet place with the documentation and read.

Performance-related faults

The family of faults that is related to performance is generally traceable to software issues, or moving even further back in the design cycle, to decisions made very early on in the life cycle of the design process where the requirements documentation

[d]https://www.ioccc.org/.

feeds into the architectural decisions. Partitioning was discussed in Chapters 3 and 6, so we won't go into any depth here, except to say that partitioning assumptions generally get tested as the system becomes more loaded, and overhead, such as the time required to do a task switch, begins to become more significant.

Fortunately, if it is software-related, or even if has to do with the choice of processor and memory decisions, these faults are relatively straightforward to find if you have the right tools. Even if all you have is an I/O port and an oscilloscope, you can still measure performance parameters. This was discussed in Chapter 1 and several times in subsequent chapters. The downside is that while you can isolate the source of the fault pretty quickly, fixing it, or alternately, bringing the system back to an acceptable performance level, might not be easy.

There are many articles concerning performance issues surrounding real-time operating systems. Most RTOS vendors offer task-aware performance tools to help you get the best performance out of your system and their software. Assuming that you don't experience system faults like the Mars Rover RTOS bug (Chapter 3), then profiling where your code is spending most of its time can easily be achieved, although deciding how to fix it may not be as simple and could involve massive amounts of redesign (time and resources) to bring the performance back up to the required performance level.

If your system includes an FPGA rather than an ASIC, then it might be possible to move a time-critical software module into the FPGA (assuming you have enough spare gates and I/O), or even just move some software into the hardware in order to reduce the loading on the processor. In any case, many of these decisions revolve around fixing what you already have, rather than doing a system redesign.

On the plus side, software can be handcrafted, although at potentially high resource penalties. Rewriting critical modules in assembly language so that they take the absolute minimum amount of CPU cycles is one solution (Chapter 1). This is the argument I use with my students when they predictably complain about having to learn assembly language. Several other chapters discuss how the choice of compiler can affect the overall performance of your code. In fact, in Chapter 6 we saw how the EEMBC benchmarks showed a 32:1 difference in performance between no optimization and a highly optimized code compilation.

Reproducible faults

This one is easy because it requires very little explanation here. Finding these faults is what this entire book is about! Perhaps I shouldn't have listed it at all because you might accuse me of doing a "bait and switch" in this section. Whether you suspect a hardware fault, a software fault, or a combination of both, the overall process doesn't change. The key is to be systematic and disciplined. Keep notes on:

- What you observed.
- The conditions that lead up to the fault.
- What you think might be causing the fault (your hypothesis).
- How you intend to test your hypothesis, including process and tools.
- What your testing has uncovered.
- What you will do next.

While this might seem like overkill, having these notes could save your butt or impress your peers when a similar bug appears in the future and you whip out your lab notebook from that time frame and point to your analysis on the page.

Here's the scenario. You are no longer a newbie engineer, but a grizzled veteran who staunchly refuses to become a generalist and join management (I saw that in a Dilbert cartoon). A new engineer comes to your cubicle and describes their problem. You first congratulate them on seeking help and when they describe their problem, you open the appropriate lab notebook and share your experience with the same or similar problem. You are God.

Intermittent faults

These are the ones that delay product releases and leave management with bald spots from pulling out their hair. The system "usually" works properly, but occasionally it doesn't. This general category of faults is not the same as the previous category where you may have a very infrequent fault, but it is reproducible nevertheless as long as you reproduce the same sequence of events. The classic example of the former is the infamous Therac-25 radiation therapy machine produced by Atomic Energy of Canada Limited in 1982 [4]. The Therac-25 gave at least six patients massive overdoses of radiation due to the removal of hardware safety interlocks (Therac-20) that were replaced with a check system based on software.

What was particularly germane here was that the safety review board report listed this particular failure mode:

The failure occurred only when a particular nonstandard sequence of keystrokes was entered...

My point is that even though the fault was not frequently observed, it was reproducible with the right sequence of keystrokes.

While we're on the subject of software glitches, we might as well continue in the software realm for a while longer. We just looked at the case study of the Therac-25, but poor code and design decisions are just one manifestation of potential sources of software glitches. I would venture to guess that today the majority of software glitches can be found in real-time systems having multiple threads or running under an RTOS, or combinations of both. All can usually be traced to interactions between the hardware, through an asynchronous interrupt and the software threads, or the interrupt service routine and the software threads. Here's a simple one. Suppose that you have two threads that are running asynchronously of each other and both can update a shared counter. These could just as easily be two CPU cores and a block shared memory. Without some blocking mechanism, such as a mutex, it would be possible for one of the threads to negate the update operation of the counter and instead of a count incremented by two, the count is only incremented by one. Depending upon the duty cycle of these events, this might be a reproducible fault or a once-every-year glitch.

In Chapter 5, I discussed how to uncover which process is clobbering a global variable. We could also list this one under the glitch category, even though it is most likely a reproducible fault. Same thing with stack overflow. We're all familiar with stack overflow as a way to hack into a computer or cause a denial-of-service fault to occur. We discussed stack overflow, how to find it, and how to avoid it in Chapter 3. However, it is worth mentioning again in this section because even though it is a deterministic and potentially predictable fault, when it can occur may be totally unpredictable and would depend upon the sequence of events leading up to the fault.

There is one reality that we must face in trying to isolate a fault in a real-time system. The reality that you might not be able to isolate it because it is nondeterministic and there is no practical way to set up a measurement that will find it. Now, admittedly, this is an extreme case, but nondeterministic faults are among the most difficult to isolate. Some glitches may be due to thermal noise or other random noise sources. This is particularly true of jitter in serial data streams. In a white paper on jitter, Keysight [5] describes the sources of jitter in a serial bit stream and how to measure it using their EZJIT software add-on package for their Infinium oscilloscope family.

Jitter can occur when noise and phase variations occur between edges of the data and the reference signals, typically a data clock. Ideally, the clock signal edge should be synchronized to the middle of the data packet at the time that the data are most stable. However, fluctuations in the relative timing between the clock edge may occur and if large enough, can cause a data bit to be misinterpreted. The white paper demonstrates a method to observe jitter by making use of the persistence function available in most modern oscilloscopes. Fig. 9.1 shows the jitter in the rising edge of a signal. The built-in measurement software enables the viewer to see the relative occurrences of the waveform over time.

According to the paper, jitter may be characterized as random jitter, as described above, or deterministic jitter, which is generally

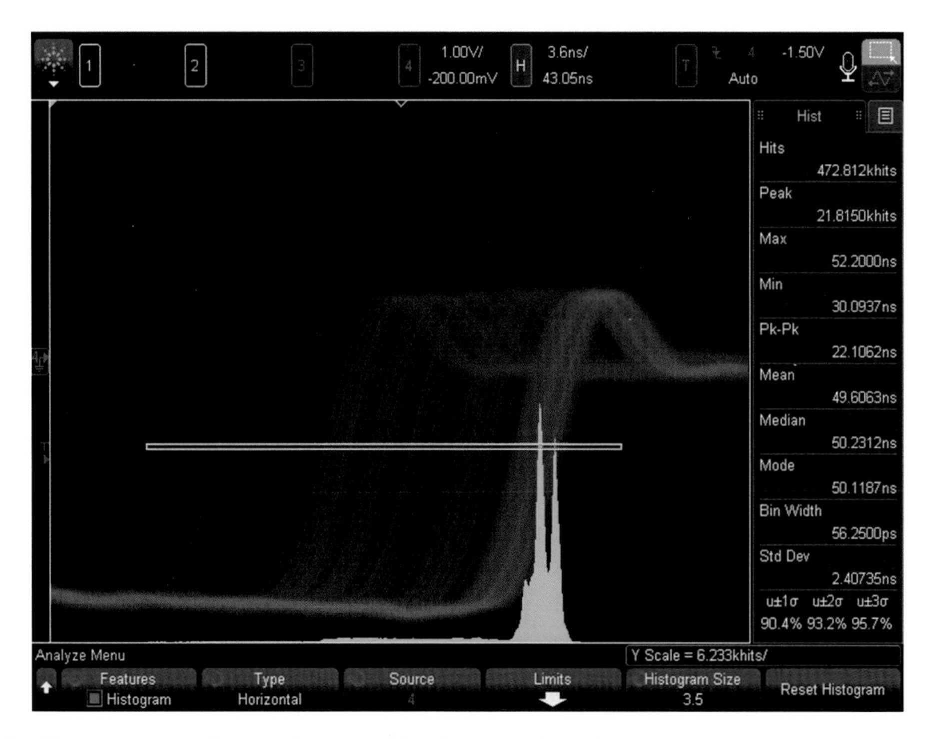

Fig. 9.1 Oscilloscope trace using persistence mode of the jitter in a clock edge. The *green* histogram is an overlay that shows the relative rate time frequency of the jitter. In this case, there are two peaks. Courtesy of Keysight Technologies.

due to design flaws or physical limitations of the components used in the design. Random jitters are most likely describable as a Gaussian distribution while deterministic jitter is often described by a bimodal distribution, as shown in the blue-green overlay in Fig. 9.1. Jitter is almost never one type or the other, and is usually composed of both elements, although one type may be dominant in a given situation. If the distribution is Gaussian, there is probably little you can do about it other than cool everything down to liquid nitrogen temperature (just kidding). However, if deterministic jitter is dominant, then you have a chance of fixing it.

In another application note, Keysight addresses the issue of crosstalk [6] and defines crosstalk this way:

> *Crosstalk is a type of distortion that comes from amplitude interference uncorrelated with data patterns. A clean signal, "victim", can be affected by crosstalk from an "aggressor" signal due to a coupling effect. The aggressor distorts the shape of the victim signal and closes the eye diagram of the victim signal.*

As engineering students, we all learned how there can be capacitive and inductive couple between signals. Depending upon the frequency component traveling on PCB traces, the spacing between the conductors, the length of the trace, and other factors, crosstalk may be deterministic if it happens frequently enough, or nondeterministic if it depends upon just the right set of conditions occurring at just the right instant of time. Fig. 9.2A and B show an "eye diagram" of the victim signal, with and without crosstalk.

Perhaps we should stop for a moment and describe an "eye diagram." The eye diagram is a popular method of displaying the quality of a digital data transmission on an oscilloscope. Imagine a long stream of data and a reference clock that may or not be embedded in the data stream itself. The oscilloscope generates the eye diagram by overlaying the data stream, all the 1 and 0s with respect to the master clock. If the system was perfect and there was no jitter or crosstalk, then the diagram would look exactly like the timing diagrams you see in a data sheet: perfect 1 to 0 and 0 to 1 transitions always occurring exactly at the same moment in time.

Fig. 9.3 is a schematic diagram that I took from a slide I use in my microprocessor class. It shows a portion of an address bus cycle with the clock signals and the all the address signals represented as a band. Because the address bus is an aggregate of many individual address lines, it is convenient to represent is as a generalized belt. The key is that all the address signals change state at exactly the same instant in time. The signal amplitudes are so

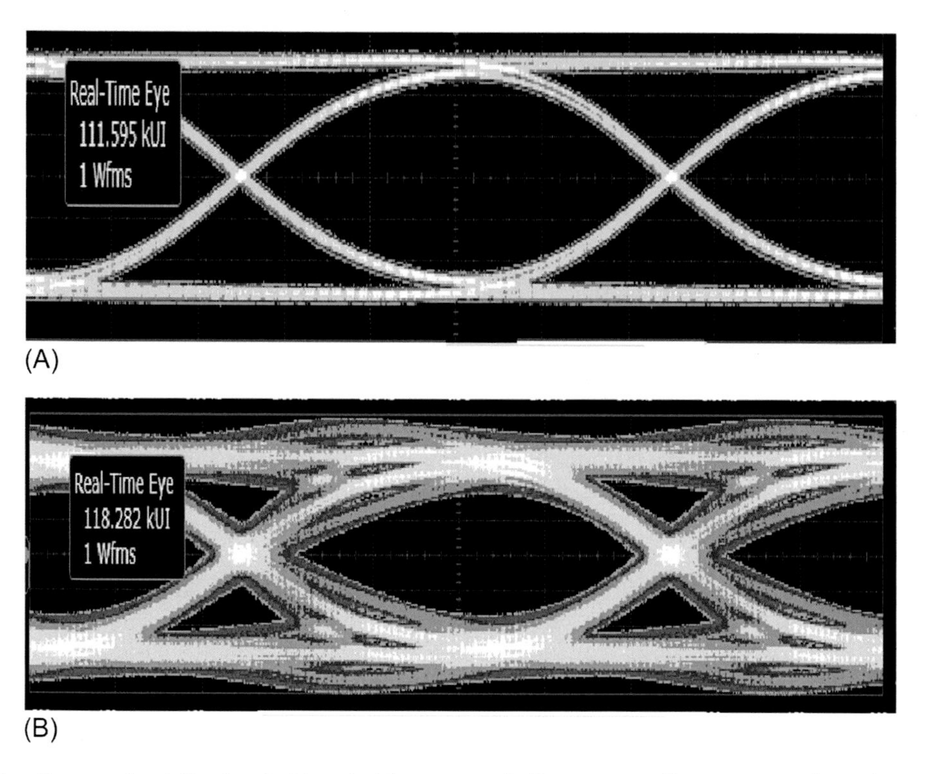

Fig. 9.2 Eye diagram of a victim signal with and without crosstalk. The upper oscilloscope trace shows the pure signal (A) and the lower trace shows the effect of crosstalk (B). Courtesy of Keysight Technologies.

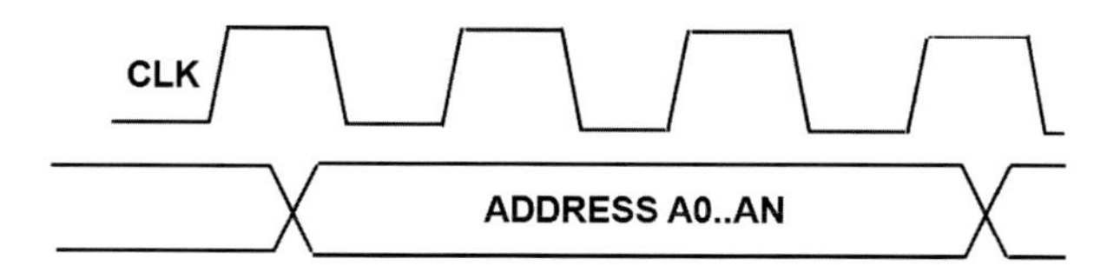

Fig. 9.3 A perfect eye diagram. All signal transitions occur at the same instant in time with the same rise and fall times and the same amplitudes. This eye diagram looks exactly like a timing diagram because all the uncertainty has been removed.

similar that they perfectly overlap and the rise times and fall times are identical. Comparing Fig. 9.2 with Fig. 9.3, we see the obvious differences between the real world and PowerPoint lecture slides.

The simplest interpretation of the eye diagram is that the more closed the eye, the more crosstalk and jitter are present.

In an article in EDN magazine, Behera et al. [7] provided a good overview of the eye diagram and how to interpret it. To quote the authors,

> *Eye diagrams provide instant visual data that engineers can use to check the signal integrity of a design and uncover problems early in the design process. Used in conjunction with other measurements such as bit-error rate, an eye diagram can help a designer predict performance and identify possible sources of problems.*

The authors show how to interpret an eye diagram. This diagram is reproduced in Fig. 9.4. We can see jitter as the width of the transition region of the low-to-high and high-to-low signal transitions and the effect of crosstalk as the width of amplitude variations. In this diagram, crosstalk is folded into a general category of signal-to-noise ratio for the signal. Clearly, we can see that the wider the eye is open, the better the signal fidelity that we can expect to have.

While the eye diagram won't find an individual glitch, it will show you the fidelity of your signal and whether this is a potentially fruitful area to continue to investigate. The other convenient aspect of using eye diagrams is that you aren't limited to just

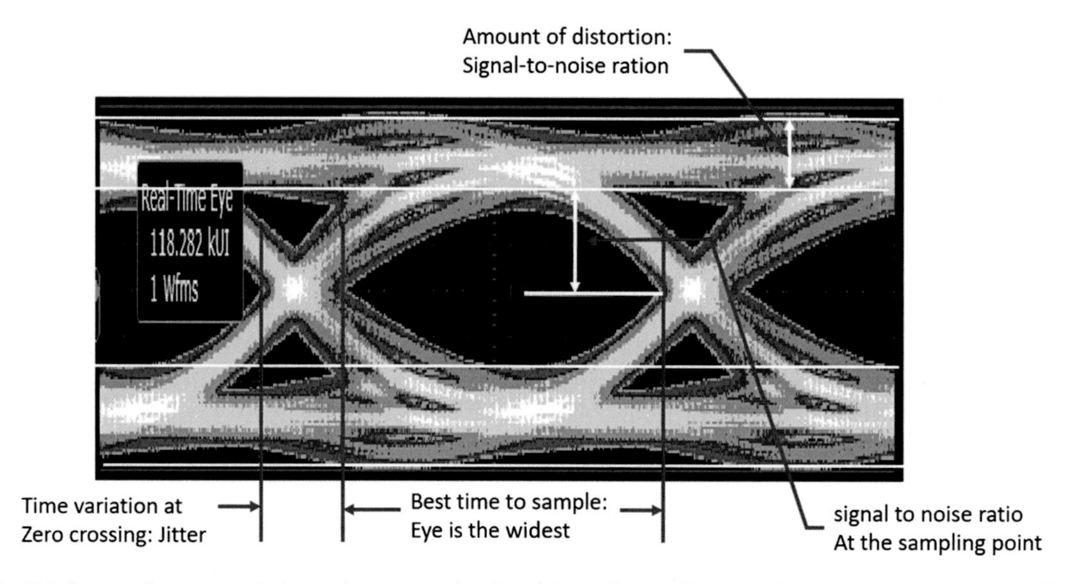

Fig. 9.4 An eye diagram can help you interpret a signal and determine the best time for making a measurement. The measuring points are taken from the Behera article and superimposed on the eye diagram from Fig. 9.3. Fig. 9.3 courtesy of Keysight Technologies.

looking at serial data streams. You can place your oscilloscope probe on any node in your circuit and collect data on signal fidelity using the eye diagram.

We digital designers tend to focus our attention to restricting our debugging efforts to the digital world and tend to shy away from analog effects, such as crosstalk. However, I hope that this discussion dispels some of that predisposition to ignore basic electronics.

Compliance faults

If your product is intended to be sold in the developed world, as I suppose most products are, then it must be designed to comply with the regulatory requirements of the country where you intend to sell the product. Simple enough, except when your product design is finished and all that is left is the regulatory testing.... Gulp. For us, the permitted level of RF emission is the most important compliance issue that we'll ever have to deal with, so this is a good one to spend time on.

Depending upon the intended use model for a product, the regulatory issues will be different as well as in the designated country of sale. Today, most products are sold worldwide, so we generally have to design the product to meet the most stringent regulatory issues in the countries that we will be dealing with. For me, when I was a hardware designer, that country was Germany.

As explained to me during one of our in-house classes on RF suppression, the German compliance agency (today it is a uniform standard within the EU) would drive a van with sensitive RF detectors around cities and town with factories in the towns. In the United States, we're accustomed to industrial plants being set away from residential areas, but in Europe, where land is very dear, a large electronics factory will butt right up against a residence.

If RF emissions are detected, the factory gets shut down until it is corrected. Of course, this only applies to the emissions from a factory as a whole, and not an individual product. I suppose that the purpose of this story was to convince the R&D engineers about the seriousness of compliance issues in the sale of products to a worldwide customer base.

I was the hardware manager for the HP 64700 family of in-circuit emulators. For cost savings, we put the emulators in plastic rather than metal cases. When it came time to test the emulators for RF compliance, they failed with several harmonics of the 16 MHz clock testing about the limit in our anechoic test chamber.

Without going into the gory details, the final "fix" cost as much as the original budget for all the rest of the hardware. We had to coat the entire inside of the plastic chassis with a conductive paint and add RF grounding sheet metal strips to the places where the plastic cover held the front and rear metal chassis and the front RF suppression shield. It gets worse. We had to add a conductive shield to our cables between the target system and the emulator. That's what happens when you don't design for RF compliance at the front end of the project.

Finding RF compliance faults is generally not something we can do in the same way that we might debug the hardware and software. It takes special instruments, equipment, and training to be able to measure and look for RF compliance faults in any computer-based system. Therefore, left's focus on some best practices as a way of preventing the designing-in of the faults, rather than fixing them later.

Here are just a few of the techniques I've used over the years. This is not intended to be a comprehensive collection of best practices, just some good payback techniques that worked for me.

Spread spectrum oscillators

I found this definition of spread spectrum oscillation on the IDT website. It expresses the problem and the technology in a really nice way, so let's use it. According to IDT [8]:

> *Electromagnetic interference (EMI) is a major challenge for designers of electronic devices. Strict guidelines enforced by the FCC and European Union regulate the amount of EMI a system can generate. Frequency references, whether crystal oscillators or silicon-based PLLs[e], can be a major source of EMI on circuit boards. Spread spectrum clocking is a technique where the clock frequency is modulated slightly to lower the peak energy generated by a clock. Spread spectrum clocking lowers clock-generated EMI from both the fundamental frequency and subsequent harmonics, thereby reducing the total system EMI.*

Another good application note is this discussion from Maxim Integrated Circuits [9]. This is very extensive and has a good historical introduction to the reasons for the FCC's introduction of FCC Part 15, which regulates RF emissions.

If your education bypassed a study of Fourier Series, or you've just forgotten it, let's do a very, very brief refresher to

[e]PLL = Phase-locked loop.

demonstrate where the problem comes from. It can be shown that for a perfect square wave (such as a clock) with a 50% duty cycle and infinitely fast rise and fall times, you can approximate the square wave to whatever level of accuracy you desire by a series of sine waves that are harmonics of the fundamental frequency. For our example, we'll use a 5V and 16MHz clock oscillator. We'll call our approximate square wave $x_T(t)$, where T is the period of the square wave.

It can be shown that:

$$x_T(t) = a_0 + \sum_{n=1}^{\infty} a_n \cos(n\omega_0 t)$$

where ω_0 is the angular frequency of the square wave, n is the nth harmonic, and a_n is the amplitude of the nth harmonic.

$$a_n = 2\frac{A}{n\pi}\left(-1^{\frac{n-1}{2}}\right)$$

when n is an odd number harmonic and $a_n = 0$ when n is an even harmonic and $n \neq 0$. Here, A is the amplitude of the square wave.

We can see that the coefficient terms of the series, represented by a_n, slowly diminish in amplitude. The following table lists the amplitudes of the first several harmonics:

n	0	1	2	3	4	5	6	7	8	9
a_n	0.5A	0.637A	0	−0.212A	0	0.127A	0	−0.091	0	0.071A
For $A = 5$V	2.5V	3.18V	0	−1.06V	0	0.635V	0	0.455V	0	0.355V
fn (16MHz)		16	32	48	64	80	96	112	128	144

From this table, we can see that even though our fundamental frequency (by today's standards) is rather low, even at the ninth harmonic, we are still producing 355mV of energy at 144MHz. We can see this in Fig. 9.5, which shows a square wave with the superposition of the approximate square wave with the first through ninth harmonics and the relative amplitude and frequency of the ninth harmonic.

The effects of the Fourier components on the RF spectrum affect the design in another, less obvious way. There is a simple rule-of-thumb equation that relates the maximum bandwidth to the rise time of a pulse. Simply put,

$$BW(\text{GHz}) = 0.35/\text{Rise time (ns)}$$

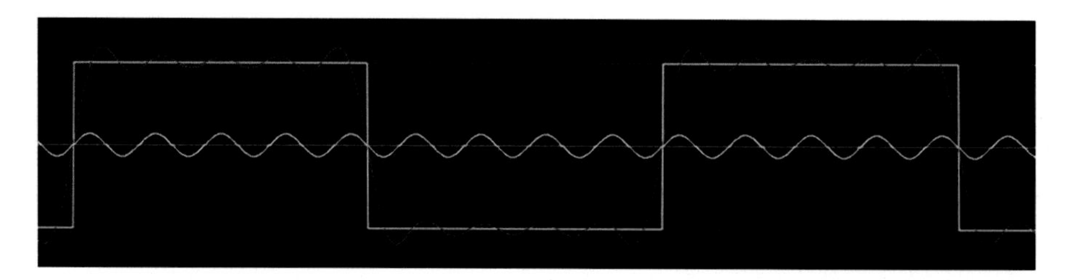

Fig. 9.5 Fourier series of a 50% duty cycle square wave *(in white)* with the Fourier Series approximation *(in red)* up to the ninth harmonic *(in yellow)*. As the higher order frequency components are added, the rise and fall times will diminish and the oscillations along the top and bottom of the wave will tend to die out. Courtesy of the author, Paul Falstad, (http://www.falstad.com/fourier/).

$$BW(\text{GHz}) = \frac{0.35}{\text{Rise time (ns)}}$$

How we define the maximum bandwidth is a bit squishy. The term is usually meant to mean the highest Fourier component of the frequency that has significant effect upon the signal. Therefore, what you mean by significant and what I mean might be different by a harmonic or two, but the takeaway is clear. One way to avoid high-frequency RF energy in your basic design is to limit the rise and fall time of the logic you are using. This will tend to manifest itself in the buffers and glue logic that you add to the design. If possible, use logic with lower amplitude (3.3 vs 5V) and slower edges (ALS families versus FCT families).

Here are some other dos and don'ts from my class lectures:

- Avoid current loops.
- Shield the clock lines on inner layers of the PC board, or run parallel guard traces.
- Avoid long clock lines and long signal runs on a board.
- Use microcontrollers and ASICs wherever possible, rather than discrete logic.
- Use RF suppression (ferrite) cores on cables.
- Shield locally, rather than the entire chassis.
- Run at the slowest acceptable clock speed.
- Terminate long traces in their characteristic impedance.
- Reverse engineer products that have solved the problem.

Thermal faults

Sorry to be the bearer of bad news, but electronic circuits get hot. Modern CMOS circuits dissipate a little bit of power every time the gate switches state. Multiple this by several billion gates and suddenly we have real power to worry about. The former senator Everett Dirkson once said, “A billion here, a billion there, pretty soon, you’re talking real money.” According to Wikipedia, “Although there is no direct record of the remark, he is believed to have made it during an appearance on ‘The Tonight Show Starring Johnny Carson’” [10].

Heat generated in an integrated circuit is a bit like that (OK, it’s a stretch, but I always liked the quote). PC gamers know this because they like to clock their CPUs faster than the part’s rated clock speed. To keep the chip from burning up, they employ serious water cooling. Server farms have to deal with major heat management issues, not just for cooling the individual servers but to get the heat out of the building.

I once visited a supercomputer manufacturer to record an interview with one of the pioneers in the supercomputer field. This was part of a project grant that I had received. To a computer hardware designer like myself, this was like going to Valhalla to interview the computer gods. I got to see their top-of-the-line supercomputer up close and personal. I was told that there was a room just the same size as the room containing the computer with nothing but refrigeration equipment in it to cool the parts of the computer. Each PCB was hermetically sealed and high-pressure coolant was sprayed directly onto each heat-generating part.

You ask, “Why am I telling you this in a book about debugging?” Good question. The reason is that we have to always think about heat, heat-generating sources, and how to dissipate the heat from our circuits.

Heat is particularly troublesome when a part gets warm enough to seriously alter its performance envelope and cause the system to fail, either regularly or intermittently. When a circuit seems to work well in simulation and then starts to behave erratically in practice, an experienced engineer will usually suspect a heat problem. Of course, a part getting hot isn’t always the primary cause of the problem, but a part getting hot is a red flag that a part failure may be down the road.

Sometimes, we actually want to use heat to our advantage. There is a classic story about HP’s first product, an audio oscillator that they sold to Walt Disney in 1940 for the movie Fantasia. The oscillator used a light bulb as a positive temperature coefficient resistor to stabilize the output amplitude of the oscillator. As the amplitude of the output increased, the resistance of the light bulb’s filament went up, which tended to drive the amplitude back down. This was a classic control loop problem.

One of my colleagues at HP kept a ready supply of Freeze Mist spray cans at his desk. Whenever he suspected a heat problem, he would spray various parts to see if the problem went away. Of course, today we would frown upon dumping all that Freon into the atmosphere, but it was an effective debugging technique nonetheless.

My preferred technique was to touch the parts with the tip of my tongue. Gross, I know, but the tongue is pretty sensitive to heat. The downside is that I burned my tongue a few times, but it did point me in the right direction.

We can see the effects of temperature on the propagation delays in data sheets. Sometimes there may be graphs depicting propagation delays versus temperature, but usually the prop delay is just listed as a range between minimum and maximum values over the operating temperature range of the part. Providing the specification in this way allows the manufacturer to fold all the process variations and temperature variations into a single specification that provides you with everything you should need to know.

However, there are rare times when you need to design something closer to the margins than you think prudent. Perhaps you decide to go with the average propagation delay rather than the maximum value. I've done that at times because the max value was way beyond what the design could tolerate. Fortunately, it never came back to bite me, but I made the decision with my eyes open to the possibility of an issue.

Unfortunately, thermal effects may not be easy to isolate because once you remove the cover from a chassis containing electronic circuitry, you are suddenly perturbing the system you are trying to measure. Taking off the cover allows the heat to dissipate more readily and the temperature in the chassis will immediately start to drop, altering the conditions of your testing.

Allegro Semiconductor [11] provides an excellent overview of the calculations related to temperature rise in integrated circuits. They even include several example problem solutions, which indicates to me that one or more of the authors of the article has academic tendencies. I strongly recommend this app note as a must-read. It has a rather straightforward set of calculations for any EE to be able to follow and use. Allegro has a nice summary list of best practices that I'm including here,

1. Modify or partition the circuit design so the IC is not required to dissipate as much power.
2. Reduce the thermal resistance of the IC by using a heat sink or forced-air cooling.

3. Reduce the ambient temperature by moving heat-producing components such as transformers and resistors away from the IC.
4. Specify a different IC with improved thermal or electrical characteristics (if available).

Allegro had another suggestion that was new to me and I found to be quite interesting. I was going to suggest that a way around the problem of removing the chassis cover was to fasten a very small thermocouple with heat conductive epoxy to the top of an IC that you suspect is overheating. Then, bring out the leads through an air vent and measure the temperature rise of the package. To quote the Allegro article,

> *The most popular technique of measuring IC temperature uses the characteristic of a diode to reduce its forward voltage with temperature. Many IC chips have some sort of accessible diode—parasitic, input protection, base-emitter junction, or output clamp. With this technique, a "sense" diode is calibrated so that forward voltage is a direct indicator of diode junction temperature. Then, current is applied to some other component on the chip to simulate operating conditions and to produce a temperature rise. Because the thermal resistance of the silicon chip is low, the temperature of the sense diode is assumed to be the same as the rest of the monolithic chip.*

The article goes on to describe the process for calibrating the sense diode by passing a control current of approximately 1 mA through the diode and then measuring the forward voltage in 25°C increments. Alternatively, the diode forward voltage drop can be measured at room temperature, followed by assuming that the forward voltage will drop at a rate of approximately 1.8 mV/°C.

At HP, we were fortunate to have mechanical engineers who understood this temperature measurement stuff and had these cute little anemometers that could measure air flow rates at various points in the chassis. These measurements were very helpful in deciding where to place heat-producing components before I started to lay out my PCB.

About a year ago, I had a student who was using a surface-mounted IC that was designed to be heatsinked to the PCB. The idea was to place a pad under the entire footprint of the body of the part, keeping it away from the pins. The pad was connected by an array of vias to the inner layer ground plane. The part was getting hot and failing intermittently. He would have to turn everything off and let it cool. Turns out that he neglected one critical part of the heat management system: thermal grease. Without a good thermal grease between the heatsink pad on the IC and the pad on the PCB, the thermal resistance was too great, and the part overheated to the point that it was failing.

Mechanical issues

Mechanical issues come about because real parts occupy space and need to be interconnected, whereas simulations are pristine. Whenever a signal enters or leaves an I/O pin of an IC, there is a mechanical interconnection somewhere in the path to the next part of the circuit. This might be a solder joint, or a socket, followed by a connector to another board, or front panel. These mechanical connections are generally benign and reliable, until they aren't. Then you can have a reproducible fault, or worse, an intermittent fault that can take a long time to isolate.

In my experience, we tend not to suspect a connection of some flavor if the fault is intermittent. If the fault is continuous, then the logical approach is to trace the signal flow until you see that the signal is missing. For some reason, if the fault is intermittent, the mechanical connection is way down on the list of possible causes. Sometimes, you can bring an intermittent mechanical fault to the top of the list of suspects when you are lucky enough to be able to correlate the failure to a mechanical shock, like we did with the HP oscilloscope that would accidently trigger when jarred (Chapter 5).

However, this is more the exception than the rule because just normal vibrations in a room or lab could trigger a delicate connection to momentarily lose contact. If you happen to own the vibration isolation tables that are used in holography, then you might be able to eliminate room vibration as a cause, but they aren't very handy or portable.

Where we've had the most problems is with insulation-displacement cables (IDC). These are the ubiquitous flat cables that typically plug into 40- or 50-pin sockets. These cables have a very limited number of insertion and removal cycles. Depending on how they are inserted and removed, this could be fewer than 10 times. Our students tend to grab the cables and pull them out of the socket, which strains the pin that pierces the cable insulation and you end up with an intermittent connection. A good rule of thumb here is to always use ejector ears on the sockets so that the cable is not strained when being removed.

Second only to the cable strain issue is the bent pin. This can happen when the cable is removed by pulling on it at an angle. This results in a bent male pin in the socket that missed the mating hole in the cable and ends up being bent to a neighboring pin, resulting in a constant fault, or an intermittent one if the bent pin touches another one.

Bent IC leads when inserting a part in a socket are another source of mechanical failure. This can be insidious because the pin can bend under and look fine on cursory inspection, and even make mechanical connections for a period of time until it starts to fail. I suspect that this failure mode is becoming less prevalent due to the emergence of surface-mounted parts. However, I advise my students that if at all possible, use through-hole parts and sockets in their initial designs in order to facilitate repairs and rework.

As previously mentioned, cold solder joints and solder bridges are a rich source of potential problems. These are relatively straightforward to find when dealing with through-hole parts, but may be much more difficult to find if the board has a preponderance of fine-pitch surface-mount parts. It may be very difficult or impossible to see a cold solder joint or a missing solder joint on a 240-pin IC, and the thought of trying to "ohm it out" won't help because as soon as you touch your probe to the pin, you push it down and make the connection.

This is the type of fault where a lot of the best practices we've discussed in this book go out the window and you have to "try stuff." As much as I hate to admit it, I've resoldered a surface-mount part hoping it was a bad solder joint. Looking back, I think I was batting about 500 overall on that defect.

Why does this problem crop up? Remember, we're not fixing defects in production boards. Our model is that you are trying to debug a circuit design that is under development and it is likely that you soldered this board yourself, or a technician soldered it as a "one-off." Until a surface-mount process is dialed in, soldering flaws are a distinct possibility.

Power supply-related faults

This is a broad category and covers a myriad of potential issues, so once again, I'll be brief. The best advice I can provide here to start out is to buy one of the standard books in the field, such as the book by Morrison [12]. This is the current edition of the book, but I've seen the third edition for sale by used book sellers for much less. I doubt that Faraday's Laws have changed very much and ground loops are still ground loops, so I suspect that either edition would be a fine reference book to add to your professional library. If you want a shorter and more concise guide to the best practices for grounding your circuit, I recommend the applications note by Zumbahlen [13] in the "Analog Devices" excellent technical journal, Analog Dialogue.

I've been a long-time subscriber to the journal and considered it to be part of my required reading. I liked it better when it was a real magazine, not an e-journal, because I could read it in the hot tub, but that's another story. Now you can subscribe to the online journal with just a few mouse clicks.

The reason I started with that recommendation is because not understanding proper grounding techniques is one of the key contributors to noise-related faults in embedded systems. If there is noise in the grounding system, that noise is going to impact the rest of the system as well. Even though your system may be all digital, noise spikes can arise and cause a bit to flip. Finding these faults is not particularly difficult. You place a scope probe on the ground rail or the power rail, set the trigger to AC coupled, set the scope for single shot, and gradually increase the vertical sensitivity until the scope triggers. Assuming your scope's bandwidth is high enough to capture the switching transients on your power and ground buses, this should work.

When you have mixed signal analog and digital circuits in near proximity, things get messier and it may be very difficult to trace the source of the problem. Usually, it manifests itself as analog readings that are outside the accuracy and reproducibility envelope that you would expect to see. So, you have a 12-bit ADC in your system, and you are only getting 8 bits of accuracy.

According to Zumbahlen,

> *It is a fact of life that digital circuitry is noisy. Saturating logic, such as TTL and CMOS, draws large, fast current spikes from its supply during switching. Logic stages, with hundreds of millivolts (or more) of noise immunity, usually have little need for high levels of supply decoupling. On the other hand, analog circuitry is quite vulnerable to noise—on both power supply rails and grounds—so it is sensible to separate analog and digital circuitry to prevent digital noise from corrupting analog performance. Such separation involves separation of both ground returns and power rails—which can be inconvenient in a mixed-signal system.*
>
> *Nevertheless, if a high-accuracy mixed-signal system is to deliver full performance, it is essential to have separate analog and digital grounds and separate power supplies. The fact that some analog circuitry will "operate" (function) from a single +5-V supply does not mean that it may optimally be operated from the same noisy +5-V supply as the microprocessor, dynamic RAM, electric fan, and other high-current devices! The analog portion must operate at full*

performance from such a supply, not just be functional. By necessity, this distinction will require very careful attention to both the supply rails and the ground interfacing.

Ideally, a mixed signal system will have a ground plane. Typically, this is an inner-layer plane of copper that provides a low-resistance and low-impedance return path for analog and digital currents. In really critical systems, as Zumbahlen points out, you may want to have your voltage delivered to the board on voltage planes rather than voltage buses. Another advantage of the ground plane is that it allows you to create microstrip transmission lines for your high-speed signals, thus providing a controlled-impedance environment for your high-speed signals and also reducing RF and EMI emissions in the process.

Students will often connect their circuit to a lab power supply using alligator leads or something like 3 ft of 22-gauge hook-up wire. Zumbahlen points out that 22-gauge wire has an inductance of about 20 nH/in. A switching logic signal can generate a current transient of 10 mA/ns. Here's where I'm going to force the students to memorize the formula:

$$\Delta V = L\frac{\Delta i}{\Delta t} = 20\,\text{nH} \times \frac{10\,\text{mA}}{\text{ns}} \times 36 = 7.2\,\text{V}$$

Yikes! Is it any wonder that they have problems getting circuits to work properly on solderless breadboards?

In Chapter 4, I recommended using an electrolytic capacitor on your board along with a liberal sprinkling of 0.1 ceramic capacitors as filters. These capacitors provide the low-pass filter function for transient noise on your power buses. Place the ceramic capacitors near the power inputs to the ICs and the electrolytic near the connector to the power supply.

I used to think that switching power supplies was a poor choice for analog circuitry, but I've slowly come around since then. I suppose it is possible to get odd effects if the switching frequency is close to a clock frequency on your board, but I've never seen the problem myself.

However, Porter [14] points out that,

Switching power supplies create their own undesired noise, usually at harmonics of the switching frequency or coherent to the switching frequency.

If you're in the least bit squeamish about using a switching supply for your analog supply rails, then use a low-dropout linear regulator between your switcher and the analog circuitry. The power loss through the regulator will be minimal and you'll have all the benefits of a low-noise power supply.

For example, Analog Devices manufactures the ADP151 Ultra Low Noise, 200 mA CMOS Linear Regulator. This part has a noise specification of 9 μV rms and a dropout voltage of 150 mV @ 200 mA load. What's not to like?[f] And the ADP151 won't bust your budget. They cost less than $0.50 each.

After writing this section, I seemed to have drifted toward a "best practices" discussion, rather than a way to find faults. If you think about it, a poorly designed board is not a simple fix, like a cold solder joint. It means a board redesign, so perhaps avoiding the fault in the first place is the best solution to the problem.

As I mentioned in an earlier chapter, be sure to include ground and power supply test points in your design. The ground test point is crucial if you want to connect an oscilloscope probe to your board and make sensitive measurements. Sometimes, a ground-referenced measurement is not possible or not desirable. A very practical Tektronix application note [15] addresses this issue with a list of measurements that don't lend themselves to single-ended measurements. These are:

- Drain to source voltage (VDS) on a MOSFET.
- Diode voltage on a freewheeling diode.
- Inductor and transformer voltages.
- Voltage drop across ungrounded resistors.

If you need to start looking into the power supply circuitry to see if it is misbehaving, then the above measurement may call for a differential measurement, either single-ended two probes and the scope set to A-B measurement, or by using a differential probe, which is the Tektronix recommended way to solve the problem.

Final thoughts

Once again, we've shrunk several volumes of how to isolate defects into one chapter. I hope you don't think that this was another "bait and switch" exercise. If it came out that way, I apologize. Ultimately, the insight for how to isolate a bug comes from experience and a deep understanding of how the system is

[f]For the record, to the best of my knowledge, I have no affiliation, financial or otherwise, with Analog Devices. I just like their parts. Before I was a digital guy, I did analog design. When I did my Ph.D. thesis, I had to measure minute changes (less than a pico-ohm) in the electrical resistance of high-purity platinum wire at 1.7 K. No kidding. That's how I learned low-noise analog design principles. Also, Analog Devices is one of the best companies for giving away sample parts, so I thank them for their generosity to my students.

supposed to work. Then, by observing the fault(s), you begin to formulate a hypothesis of what could be the cause, then move on to more measurements and either eliminate a possible cause or just dig deeper as the reason for the fault begins to reveal itself. Along the way, you still need to practice the good investigative techniques that were discussed in the preceding chapters. Above all, keep a record of what you observe, what you think is happening, and what you propose to do as a result.

Understand your circuit and your code.[g] Understand your tools and how to use them to your advantage. As a last resort, read the user's manual.

Ask someone else. Sometimes just explaining the problem to someone else will give you a new insight that you can go test. When I was on a sabbatical leave of absence at the University of Sydney, I was working late one night and I couldn't get an LTspice simulation to work properly. I had set a goal for myself that I would get the circuit working and then go home. I left my office and just walked around thinking. I ran into another faculty member (a database person) who knew absolutely nothing about circuits, and I asked him if I could explain my problem to him. He said sure (I assume because he was ready to leave) and I proceeded to show him the simulation I was running and the crazy behavior I was observing. As I explained it, I suddenly had an idea of what it could be and tried it out. That fixed it! I told him he was an analog wizard and we both went home happy.

Call the vendor and speak to an applications engineer. This is particularly true if the bug involves a part that is new to you. Datasheets are not textbooks. Not everything is spelled out in great detail with example problems at the end of the datasheet. I tell my students that the important result is a solved problem, with or without outside assistance. As long as you give credit where credit is due, you are a good engineer.

We all like to believe that ultimately, these systems are deterministic. There is a cause and that results in an observable effect. Sometimes the cause is obscure and will take real detective work to find. Imagine that you are inventing industrial controllers and your controller might have to be used in a metal foundry that employs electric arc furnaces to melt the scrap metal. When that arc strikes in the metal, several kiloamps of current suddenly begin to flow and the EMI pulse that is generated can be very, very large. Unless you actually test for this level of EMI sensitivity,

[g]Of course, sometimes you are at a vendor's mercy and the fault lies further up the food chain. You did nothing wrong. The bug was a typo on a data sheet, or a bug in a library routine that you bought or inherited.

you'll never see the fault until you test it under the conditions in this foundry, or one similar.

I remember a Dilbert cartoon where the pointy-haired boss is having a computer problem and Dilbert goes to help him. The boss keeps hitting the same key over and over again. Dilbert comments something to the effect that the definition of insanity is to keep doing the same thing over and over again and expect to get a different result. Suddenly, the boss hits the same key again and it starts to work. Dilbert comments that this is setting a very troubling precedent.

Software engineers see similar problems with their compilers. They know that when they get the message that there was a compiler error at line number 489, that is probably the place where the compiler could not resolve the error that actually occurred many lines before.

So, take heart. The same superstitions that you follow with your favorite sports team when they are on a winning streak "probably" will not help you find and/or fix the bug. Maybe the ripe odor you get from not changing your socks for a week will help you think about the problem in a new light. Who knows?

What I can say is that when you have a low-duty cycle bug (perhaps once a week, once a month, etc.) the best approach is to try to make it a high-duty cycle bug (once a second). Then you have a potential clue to the cause (what you did) and that will lead to insight.

In Chapter 3, I noted a problem we were having with a software performance analyzer (SPA). The bug was ultimately traced to a typo in a data sheet, but it occurred very infrequently, but reproducibly. So, we figured out a way to increase its frequency by making the Gray Code time stamp clock with fewer bits. This caused the problem to occur really frequently and we could see the problem on a scope. The solution was pretty straightforward after that.

Onward!

References

[1] https://www.pcmag.com/encyclopedia/term/43032/fault-isolation.

[2] Take the Mystery Out of Probing: 7 Common Oscilloscope Probing Pitfalls to Avoid, Keysight Technologies, Inc, 2018. eBook number 5992-2848EN, Published March 20.

[3] https://siglentna.com/operating-tip/determine-bandwidth-scope-require-application/.

[4] N.G. Leveson, C.S. Turner, An investigation of the Therac-25 accidents, IEEE Comput. 26 (7) (July 1993) 18–41.

[5] Keysight Technologies, Inc, Left Turn or Lake Front: Understanding and Measuring Jitter: a white paper, 5992-3560EN, December, 2018.

[6] Keysight Technologies, Inc, Application Note: Overcoming Crosstalk Challenges in Today's Digital and Wireless Designs, Published in USA, 5992-1610EN, March 7, 2019.
[7] D. Behera, S. Varshney, S. Srivastava, S. Tiwari, Eye Diagram Basics: Reading and Applying Eye Diagrams, EDN network, December 16, https://www.edn.com/design/test-and-measurement/4389368/Eye-Diagram-Basics-Reading-and-applying-eye-diagrams, 2011.
[8] https://www.idt.com/products/clocks-timing/application-specific-clocks/spread-spectrum-clocks?utm_source=google&utm_medium=cpc&utm_campaign=timing&utm_content=sscg&gclid=EAIaIQobChMIg9aR74Os5QIVhBx9Ch2aUgkzEAAYASAAEgLmnfD_BwE.
[9] https://www.maximintegrated.com/en/design/technical-documents/app-notes/1/1995.html.
[10] https://en.wikipedia.org/wiki/Everett_Dirksen.
[11] Allegro Semiconductor Corporation, Application Note 29501.4, Reprinted by permission of Machine Design June 9, 1977. Issue, Copyright © 1977 by Penton/IPC Inc., Cleveland, Ohio.
[12] R. Morrison, Grounding and Shielding Techniques in Instrumentation, sixth ed., Wiley-IEEE Press, 2016. ISBN: 978-1-119-18375-4.
[13] H. Zumbahlen, Staying Well Grounded, Analog Devices, Analog Dialogue, 46-06, June, www.analog.com/analogdialogue, 2012.
[14] A. Porter, Tips and techniques for power supply noise measurements, EE Times (2006). December 11, https://www.eetimes.com/document.asp?doc_id=1273143&page_number=1.
[15] Tektronix Corporation, Probing Techniques for Accurate Voltage Measurements on Power Converters with Oscilloscopes: Application Note, 2016.

10

Debugging real-time operating systems (RTOS)

Introduction

We're all familiar with our PC-based operating systems. Windows, for example is the most common operating system in the world today, although we might argue that the Android operating system is giving Windows' dominance a serious challenge. On the Apple side of the PC and smart phone wars, we have the iPhone and MacOS flavors as well as variants for other Apple products.

For the purists and software developers, there is Linux and its father, UNIX. Going even further back, we have the operating systems developed for the mainframe and minicomputers, as well as the millions of different operating systems written by computer science students for their O/S class projects.

Real-time operating systems are designed to provide the advantages of the computer operating system to the world of real-time, deadline-driven applications. Due to the demands of "real time," the operating system has a tough row to hoe. It must provide all the usual services and advantages associated with operating systems, but it must also deal with the reality that all the applications that may be running at any given moment in time are not equal in importance. Some applications, or tasks, are more important, and must be given priority over tasks that are less important.

This prioritization of tasks is what differentiates RTOS-based systems from systems based upon the more common, round-robin system, where each task gets a slice of time, and the tasks are basically allocated equal time slices. Of course, there are variants here as well. I remember in earlier versions of Windows, a task could become unresponsive and slow down if there was no user activity in the window after some amount of time had elapsed. The workaround was to periodically click the mouse in the window to "bump" the program back to life.

Debugging Embedded and Real-Time Systems. https://doi.org/10.1016/B978-0-12-817811-9.00010-7

Much of this chapter will discuss the types of bugs that typically arise in RTOS-based systems as well as methods of preventing them and finding them. However, once we discuss the unique issues that are associated with these systems, what remains is the same methodologies and tools that we discussed in previous chapters. These include:

- Keeping notes on observations.
- Understanding the capabilities of your tools and how to use them.
- Having the right tool for the problem you are trying to solve.
- Seeking out the wisdom of others.

Defect issues in RTOS systems

An excellent starting point for our discussion is an article in the online journal, embedded.com, called *Real-time debugging 101* [1]. Let's use this article as our jumping-off point for a discussion of bugs associated with RTOS systems. The authors of the article are quick to point out that the historical methods of debugging, such as printf() statements and blinking an LED to indicate a point in the program has been reached won't work in an RTOS environment, and they point out the obvious reasons why this is so. Other standard debugging techniques such as setting breakpoints and single stepping are possible if the debugger you are using is RTOS-aware, but there may be side effects that are as hard to uncover as the original problem.

The article classifies the most commonly found problems into several general categories:

- Synchronization problems.
- Memory corruption.
- Interrupt-related problems.
- Unintended compiler optimizations.
- Exceptions.

Let's follow embedded.com's lead and look at each of these areas and issues in turn.

Synchronization problems

Synchronization problems can exist in any system, RTOS or not, where you have real-world events that are inherently not synchronized to the system clock or to each other (asynchronous) sharing a resource in your system. This could be a peripheral device, or a common memory location that is being used as a mailbox for message passing. When one task, thread, or interrupt is accessing this resource, and then another task, or perhaps a higher-level interrupt, takes over and modifies the common resource before the first task completes, then there is a high probability that the data will be corrupted.

This is shown in Fig. 10.1, below. Thread-A is active, and it is going to increment a memory location called "counter," which is storing a count value. Because a memory location can't be automatically incremented or decremented without the use of the ALU, the current count value must be brought into a local register, incremented, and then written back to memory. This would take three instructions in assembly language, unless the processor had a single assembly language instruction for incrementing or decrementing memory.

Even a single assembly language instruction that increments or decrements a memory variable would require a read-modify-write back operation, but at least it could not be interrupted because instructions are considered to be atomic and any interrupt must wait until the instruction completes.

Returning to Fig. 10.1, while Thread A is modifying the counter, Thread B asynchronously wakes up, reads the counter variable into Reg 2, decrements Reg 2, and then writes the new value back to the counter variable.

Thread B now goes back to sleep or moves on to other things, content that it has done its job well. Unfortunately, Thread A still

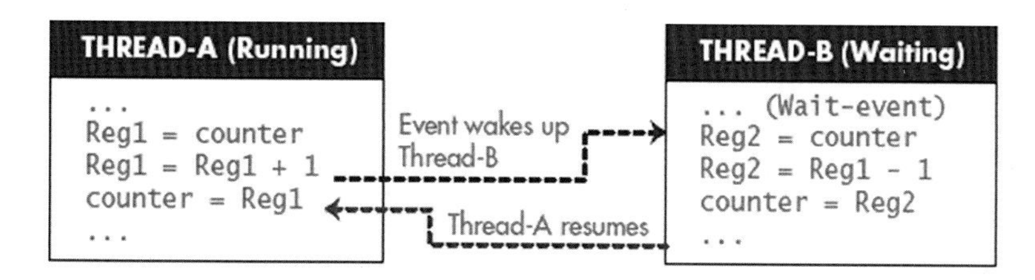

Time	Thread-A	Thread-B	Counter
0	Reg1 = counter Reg1 = Reg1 + 1 [Reg1 → 3]	waiting	2 (initial)
1	Waiting	Reg2 = counter[Reg2 2] Reg2 = Reg2-1 Counter = Reg2 [Reg2 1]	1
2	Counter = reg1 [Reg1→3]		3

Fig. 10.1 Schematic illustration of a thread-to-thread synchronization problem. Thread A and thread B share the resource called "counter." Courtesy of embedded.com.

hasn't finished, and its last operation is to write the incremented value back into the counter variable, thus negating the decrement operation of Thread B by overwriting the counter variable.

This is an insidious kind of bug because it can happen randomly and even worse, very infrequently. All you need are two threads with vastly different duty cycles, so the probability of this data corruption was extremely low, but not infinitely low. It's the kind of bug that drives software engineers to leave technology and become organic farmers.[a]

The way to avoid this kind of problem is to first realize that Murphy's Law[b] will take over in this situation, so you must guard against this from happening. Even if the system is not using an RTOS, you can turn off interrupts until the Thread A instructions complete. This has a downside, as of course it must, and will result in additional overhead for the interrupt associated with Thread A.

Memory corruption

If both threads are part of an RTOS, then the O/S usually contains mechanisms for protecting shared data. It is up to the designer to choose the most appropriate one to use. In this particular example, a mutex, or mutually exclusive locking mechanism, might be appropriate. Another method is semaphores, and you can get into a religious war debating whether a binary semaphore and a mutex are the same,[c] but as a hardware designer by trade, my eyes glaze over when these discussions start. Suffice it to say that this bug can be avoided, or fixed, once you've found it.

Fortunately, a logic analyzer is an ideal tool for finding a memory variable corruption bug because "if" you can observe the memory location, you can selectively record all accesses to it, not just trigger the LA to fill the buffer on the first access. HP (now Keysight) logic analyzers could do this, and I'm confident that those from other manufacturers can do this as well.

Depending upon the frequency of the bug, you may not want to capture every access to the variable, but only those accesses that are defined by more than one memory write operation to the variable in some window of time. For example, suppose that Thread A occurs once a second and Thread B occurs once an hour. Any event that is a write to the counter variable followed by

[a]Nothing wrong with being an organic farmer and I shop at Whole Foods like everyone else.

[b]If something can go wrong, it will.

[c]For a reasonably readable discussion, see https://www.geeksforgeeks.org/mutex-vs-semaphore/.

another write operation in less than 1 ms would definitely be suspicious, and you can set up the LA trigger to capture that event.

The problem here is whether the counter variable is cached on the CPU, or else saved in a register by a compiler optimization. You would have to make sure that the counter variable is protected, or declared using the keyword "volatile." This forces the compiler to leave it alone and not try to optimize it. Compiler issues will be discussed later in this chapter.

It is rare for an embedded system to employ hardware-enabled memory management through an MMU, even if the MMU is available in the processor. If an MMU is used, you could enable memory protection and possibly prevent memory corruption from occurring. The problem with turning on the MMU is that you suffer a performance penalty due to MMU overhead. This could occur when the processor is running in user mode (memory is protected) and it must switch to supervisor mode (memory is unprotected). With an RTOS in place, the mode switching will occur many times as the RTOS does its work, thus suffering the mode switch overhead each time the RTOS takes over.

In Chapter 3, we discussed the issues associated with stack overflow and RTOSs. How big the stack should be for your application quite naturally depends upon the needs of the RTOS and the complexity of the algorithms. If memory is plentiful, then make the stack really huge and pare it back as you exercise the code. As pointed out in Chapter 3, the simple test is to fill the stack with a known pattern and then see how big the actual stack grows (water line) while the code runs long enough to give you a good read on the real size requirements. Add 10% to be safe and you've prevented a potential bug.

One good general rule is to use large local variables with some caution because local variables are passed and live on the stack. And of course, avoid reentrant code, no matter how elegant it makes the source code appear to your buddies.

Interrupt-related problems

Interrupt service routines, or ISRs, are often written in assembly language for both speed and the ability to directly manipulate the hardware. Also, in my experience, they are usually written by the hardware designers rather than the software designers, so the potential is always there for a miscommunication or a bug. I'm not saying that EEs are incapable of writing a decent ISR, but software just isn't our thing. The closest we like to get to software is writing

Verilog code, and even that takes most of the fun out of it. But I digress.

It is easy for an ISR and the RTOS to get into each other's way. Most processors allow us to set an interrupt priority level, so that a higher-level interrupt can take over from a lower-priority interrupt. If the lower-priority interrupt becomes inactive due to another interrupt, and the lower priority interrupt is using a system resource, this may effectively block other RTOS tasks from being able to run.

Another issue with an ISR written in assembly language is that, in the immortal words of Leonardo DiCaprio in the motion picture *Titanic,* "I'm king of the world!" Once you enter assembly language you can literally do anything you want, particularly if there is no distinction made between user mode and supervisor mode because supervisor-level instructions are now freely available, thus removing another layer of protection from the code.

Referring back to Fig. 10.1, we saw how a synchronization problem can lead to the corruption of a register. The same can be said for an improperly written ISR in assembly language. The issue here is that the ISR is asynchronous and may occur at any time. If a thread is executing at the RTOS level, it is possible for an ISR to come in and change the value of the register being used by the thread.

You may argue here and say that an ISR should, on entry, save the registers being used and restore them on exit. Wouldn't that eliminate the bug? This is correct, assuming that the ISR is correctly written. If a register is used by the ISR and that register also happens to be used by the RTOS, and that register is not saved and restored, you have a bug waiting to happen. It is also important to save and restore the status register, or registers. This is a more subtle bug, but no less dangerous.

Suppose that one of the RTOS threads is executing a loop where it is counting down from some value to zero. When it hits zero, the processor should exit the loop and the code should move on. That's the theory. I'll write a little pseudo-assembly language loop to demonstrate the problem:

Line	**Label**	**Instruction**	**Comment**
1	Loop	DECR #1,R1	Decrement Register R1 by 1
2		BNE loop	If $R1 \neq 0$, go back to label "loop"

We know that ISRs cannot interrupt an executing instruction, so when the instruction in line 1 executes the last time through the loop when $<R1> - 1 = 0$, the Zero Bit in the status register will be set to 1 (result is zero).[d] The instruction at line 2 is a mnemonic for branch not equal. The branch back to "loop" will not be taken if $Z=1$ (result was zero).

Of course, this is the spot where Murphy steps in and the interrupt is taken immediately after the DECR instruction completes, but before the BNE instruction can test the value of the Z bit.

If the status register is not saved and restored, then the ISR will likely change the value of the Z bit back to 0, the count value in R1 will underflow, and the loop will continue to run.

In order to see if this scenario could be the cause of the bug, you might try to output the value in R1 through a printf() or cout() function, but we're dealing with interrupts here, and these printing functions are major perturbations to a real-time system. Once again, the logic analyzer could be the tool of choice. You could approach this in a number of ways. For example, you could add an assembly language instruction to write the value in R1 out to a noncached memory location and have the LA watch that memory address. We could call this a low-intrusion printf(). Here, your overhead would be a single instruction, but even that may not be acceptable.

If you are willing to disable the instruction cache, you could watch the instruction flow and set the trigger of the LA on the entry to the loop and trigger the LA on an instruction fetch from an address that is not the address of the two instructions in the loop. This would show you two things: (1) Did an interrupt occur while in the loop, and (2) did the loop exit properly?

Assembly language is also used when it is necessary to write the tightest possible code with no overhead due to the normal conventions of the high-level language being used for the less time-critical routines. In these cases, parameter passing between the two languages becomes an issue, and unfortunately, there is no fixed rule for how to do this. It is compiler-dependent. So, if you are used to passing parameters in back to a C function in R8 and this compiler uses R16, you have a bug, unless of course you are the type who always refers to the 500-page compiler user's manual before you code the assembly language routine. Likewise when going in the other direction from the C function to the assembly language function.

While we're on the topic of making sure that your ISR plays nice with your RTOS, I found these two rules that I thought are quite useful to point out. Quoting from the course lecture notes [2],

[d]This always drives my students crazy when I try to explain that the 0 bit gets set to 1 when the result is 0. I'm sympathetic to their pain, and my excuse is that I'm just the messenger.

1. An interrupt routine must not call any RTOS functions that might block.
   a. Could block the highest priority task
   b. Might not reset the hardware or allow further interrupts
2. An interrupt routine must not call any RTOS function that might cause the RTOS to switch tasks.
   a. Causing a higher priority task to run may cause the interrupt routine to take a very long time to complete.

One of the ramifications of these rules is that your software and hardware folk cannot write their code in isolation of each other. While your software team can usually write their parts of the code in relative isolation, the ISR or other assembly code lives in the space between the hardware and the rest of software, including the RTOS, just one level removed from the hardware. As such, it sits in a privileged position and can run roughshod over all the other rules and protections provided by the compiler and the RTOS.

What you want to avoid is finger pointing. No comments such as "It's your fault" are permitted. Here's where well-enforced design rules become worth the time it took to create them. Make sure that the hardware team understands the structure and code conventions that are required by the RTOS. If you don't want to aggravate them any more than is absolutely necessary, you can just make it a point to have the software team carefully review any code written by the hardware team and make sure it is up to your standards. Depending upon how sensitive the designer, you might make the fixes yourself, or point it out during a formal code inspection. I would prefer the code inspection because it is a nonthreatening environment.

Unintended compiler optimizations

High-level languages grudgingly accept the fact that they're not always running on a mainframe and have to be able to deal with messy hardware. This is where the "volatile" keyword comes into play. But sometimes declaring a memory operation as volatile isn't enough. A very typical situation is when the compiler, in its quest to make the code as efficient as possible, reorders some block of instructions in order to take advantage of some obvious parallelisms in the code. For example, compilers love instructions that initialize a register with a literal value. They can put that instruction pretty much anywhere and not break a sweat.

The real problem arises when it tries to reorder instructions that deal with hardware and should not be reordered. For example, FLASH memories can have a specific algorithm for writing to the memory, and the code must be executed in an exact order. NXP Semiconductors [3] offers a tutorial on how to program the FLASH memory in its 68HC08 family of microcontrollers. According to the tutorial, writing to the FLASH memory is a 13-step process involving a number of time-critical steps and dependencies on the order of execution. While it is unlikely that you would program this entirely in C, you might be tempted to do just that. If the compiler sees that there are reordering opportunities, your programming algorithm will not work, even though you coded it perfectly.

According to the embedded article, one clue to this occurrence is if the code runs correctly with the debugger code enabled, but fails when the release version of the code is compiled and built. Here's where you need to step in and do some micromanaging of the compiler to make sure that certain functions are not to be reordered. This would be preferable to turning down the optimizations for everything and suffering the ensuing performance hit.

Exceptions

Exceptions are usually unintended events that occur during processor operation and are synchronous with the processor clock and program execution. Exceptions can also be user generated. The classic exception is the famous BIOS calls in the original CP/M operating system, which evolved to CP/M-86 and finally Windows.

Two of the most common exceptions are a divide by zero error or an illegal instruction. In the case of the illegal instruction, it could be a program trying to use a supervisor-level instruction in user mode or an illegal op-code, which is generally caused by an errant pointer that drops the program counter off in the middle of nowhere, and the instruction decoder tries to make sense of the code it sees.

The venerable 68000 processor has an exception handler that is easy to understand, so we'll use this as our example. While most processors share some common exceptions, many are specific to the system architecture and are unique to the particular processor family. The 68000 (68K) family handles exceptions through an exception vector table at the beginning of memory.

Each of the first 256 long word (32-bit) memory locations can hold a 32-bit long address. Thus, the 68K supports 256 pointers to memory that will allow the programmer to write code to deal with these exceptions when they occur, and allow the system to recover gracefully.

Vector number	Address (Hex)	Assignment
0	00000000	RESET: Initialize supervisor stack pointer
1	00000004	RESET: Initialize program counter
2	00000008	Bus error
3	0000000C	Address error
4	00000010	Illegal instruction
5	00000014	Zero divide
6	00000018	CHE instruction (Check for array bounds check)
7	0000001C	TRAPV instruction (Exception if overflow bit V is set)
8	00000020	Privilege violation

Fig. 10.2 Table of the first nine exception vectors for the 68,000 processor. Each of the addresses shown contains the address of the first instruction of the user-created error handler.

When an exception occurs, the processor automatically begins the exception processing cycle and, depending upon the type of exception, will move the appropriate vector (memory address) into the processor's program counter, where the next instruction will be fetched from that address.

Fig. 10.2 is a table of the first few exception vectors for the 68 K family. Note that the exception vector table contains a heterogeneous set of vectors. That is, the table contains a mixed set of vectors that cover exiting from a RESET state, run-time problems, interrupt service routines, and user-generated interrupts (trap calls). The trap calls are generally used to enter the O/S.

When a system is running under an RTOS, the actual exception vectors will typically lead to entry points into the RTOS, and it is the responsibility of the RTOS to handle them. There may be some user-specific coding that would need to be done in cases where drivers may need to be written for low-level error handling.

With embedded systems, most exceptions will be errant pointers, stack overflow, or the kind of register corruption that we've previously discussed. These errors are pretty dramatic and can usually be tracked down by examining a trace of the program flow. If you want to do this in real time, you'll need to trigger the logic analyzer at the end of the trace buffer, not the beginning.

The reason is that the exception vector is likely to be an entry point into the RTOS and at that point, the code will become invisible to you. Tracing what happened up to the generation of the exception will usually point to where the fault occurred, providing that you have visibility into the instruction code flow. This would mean running with the I-Cache turned off, or with some hardware trace assistance built into the processor.[e]

[e]https://nexus5001.org/.

Fortunately, most processors designed for use in embedded systems contain a debug core of some kind as well as special registers to help in the debug process. Many contain trace buffers that would enable you to track the program flow up to the exception point. However, this leads to an obvious question that should be asked at the very point in your project when you are about to choose a processor:

How does this processor's debug capability support my application?

It isn't enough to simply consider benchmark performance as your acceptance criteria. You need to think about the back end of the project when time is short, the stress levels are high, the big trade show is weeks away, and you can't find the !@#$!#@ bug that has got you dead in the water.

RTOS-aware tools: An example

The processor's debug capability is not just the hardware support built into the core. It is that part of your entire tool chain that is devoted to real-time and RTOS debugging. So, using this question as our entry vector, let's examine how RTOS-aware debug tools can aid in your debug processes.

I'm going to focus on the tools offered by Lauterbach GmbH.[f] I chose Lauterbach for several reasons, but as a disclaimer, I don't have any financial arrangements with the company. First, they've been a player in the embedded systems debug space for as long as I can remember. Their support matrix for RTOS debugging and processor support is very broad, and finally, they were quite open and supportive of the creation of this book.

In 2009, Lauterbach became the world's largest provider of microprocessor development tools. I learned this from their history page. The company is still run today by Stephan Lauterbach, the brother of the founder, Lothar. Of personal interest to me is that the company began in 1979, which was the same year that I joined the Hewlett-Packard division in Colorado Springs. As I looked through Lauterbach's "history" slide show, I was struck between the parallels between the evolution of their product offerings and those of the HP Division I worked for, although I think their evolving focus has more and more focused on tracing program execution flow under various conditions and system architectures. I guess this is another reason that I find symbiosis with their tool chains. I believe in that value of real-time trace to solve real-time debugging issues.

[f]https://www.lauterbach.com/frames.html?home.html.

So, if I were having to choose a tool chain, I would certainly consider that my vendor/partner has been around for a while and whether they support the processor and RTOS I intend to use. With more than 100 processor architectures supported and more RTOS than you might ever consider, including all the heavyweights, Lauterbach would have to be a serious contender for my business.

Of course, there are other factors that will be part of your tool decision matrix. I've not considered costs, support, training, documentation, and other factors that might be considered to be part of the "complete solution."

I have no doubt that other vendors all offer various tools that overlap with Lauterbach and with each other's offerings, and sometimes the deciding factor ends up being who has the best swag to give away at the Embedded Systems Conference.[g]

So, I hope that without any loss of generality, let's look at the tools Lauterbach offers for debugging embedded RTOS-based systems. The Lauterbach debugging tools are built around their Trace32 hardware-assisted debug system. This is a modular system with insertable components. Trace32 is shown in Fig. 10.3.

According to their web page on RTOS-aware debugging [4]:

> *The TRACE32 RTOS Debugger is an adaptive debugger, which allows debugging on target systems using real-time kernels. The debugger is fully integrated in the user interface. It allows the display of kernel resources, task selective debugging and many sophisticated real time analysis functions. The analysis functions include symbolic system call trace and detailed performance analysis functions.*
>
> *The configuration to different RTOSes is controlled by a dynamically loaded extension. By changing this extension, the user may adapt the debugger to nearly any RTOS. Standard configurations are available for the most used kernels. The extension can also be used to define user specific windows for any kind of special data structures. The supported features vary between different kernels. Not all features are supported for all processors and kernels.*

In order for any hardware-assisted RTOS tool to work with another vendor's RTOS, there needs to be some level of cooperation between the two companies. It is usually mutually

[g] I hope that's not really the reason, but it might improve the overall quality of the giveaways.

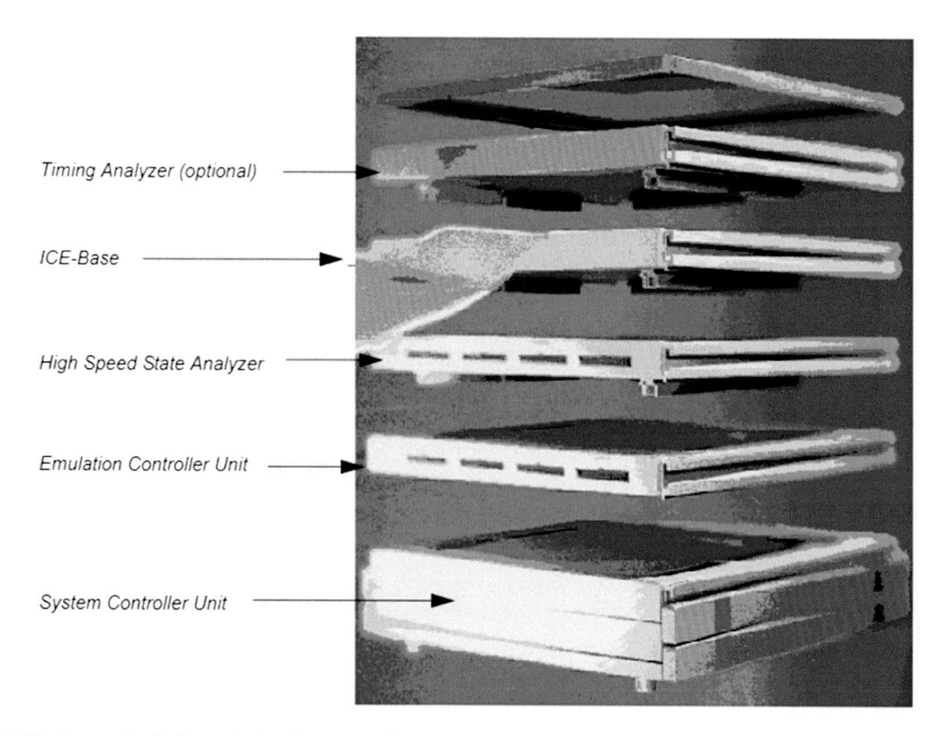

Fig. 10.3 The Lauterbach Trace32 hardware-assisted debug modular architecture. Subsystems may be added to implement run-control, overlay memory, real-time trace, and performance analysis for a wide range of processor architectures and real-time operating systems. Courtesy of Lauterbach, GMBH.

beneficial to have these cooperative agreements. Having hardware-assisted debugging support makes the RTOS more attractive to customers with hard real-time applications and being able to offer a broad range of support for many different RTOS packages makes the tool vendor's product line more compelling to prospective buyers.

A simple example of this cooperation could simply be that the RTOS kernel would do a data-write to a memory location that the debugging tool was waiting for. In our simple example, this write operation would coincide with each time the kernel initiated a switch to another task. The data would contain information that would identify the task, and perhaps other dynamic information about the current RTOS environment.

Through nondisclosure agreements, both companies can work together to improve the synergy of their products. Difficulties arise when one vendor has products that compete with the potential partners. This was the case in the early days of in-circuit emulation. One of the silicon manufacturers was also a debug tool vendor for their silicon products. Special debugging hooks put into the silicon were kept from other tool vendors, giving the silicon vendors a big advantage in the market.

The Nexus 5001 On-Chip Debug Consortium (See footnote e) addressed this issue in a very novel way. The standard allowed for special messages between Nexus-compatible silicon and tools. A tool would ignore a special message and not complain (go belly-up) if it didn't understand one of these special messages. However, vendors who had entered into these cooperative agreements with the silicon manufacturer would know what these messages mean, and their debugger interface would be able to use the information for the engineer trying to debug their code.

Fig. 10.4 shows a very detailed (and very busy) screen image of an RTOS-aware debugger. The information provided in the many tables would greatly facilitate the ability of an engineer to quickly

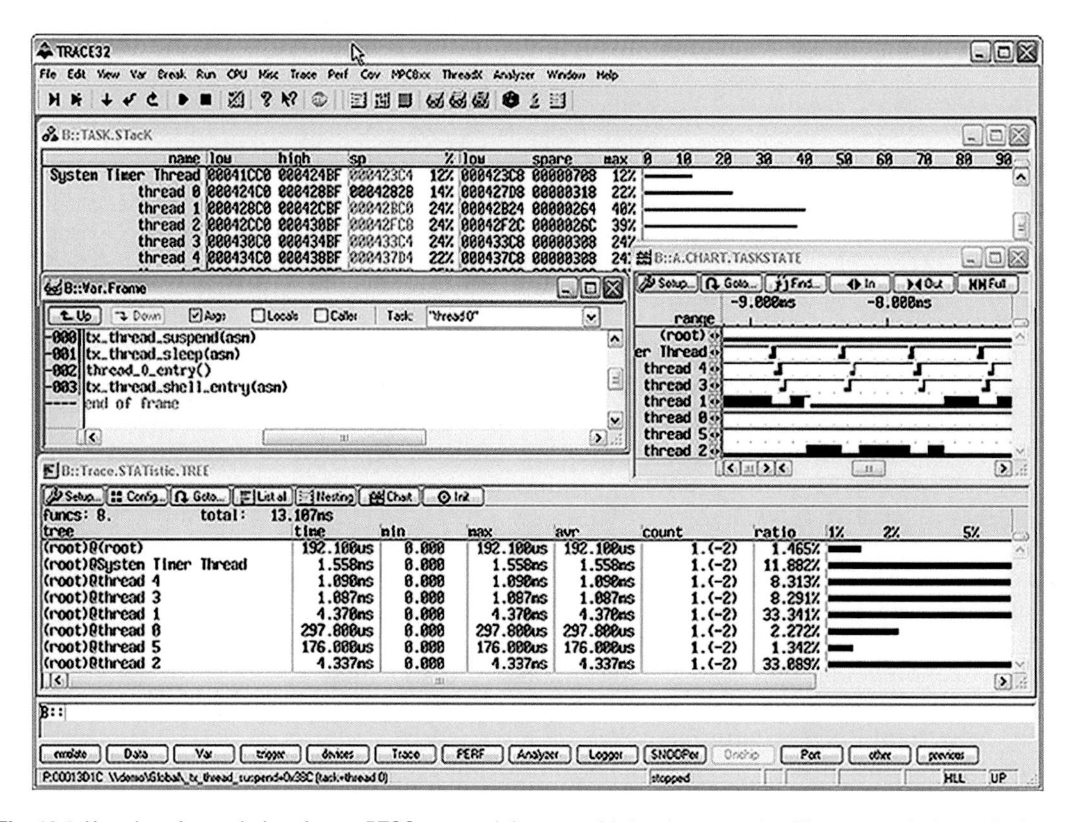

Fig. 10.4 User interface window for an RTOS-aware debugger with hardware assist. The many windows display information relating to kernel resources, CPU task usage, performance analysis, task states, and other statistics related to the behavior of the system. Courtesy of Lauterbach, GMBH.

find and fix RTOS-related problems, or for that matter, problems with the code for a particular task that only become visible with the system running at full speed.

Of particular relevance is the display's information on kernel resources. An engineer can see: [4].

- Tasks
- Queues/semaphores
- Memory usage
- Mailboxes

From my perspective, insight into processor performance is a key dynamic metric to be able to follow and document. Here, the user can view:

- Time spent in a task.
- Number of times a task switch has occurred.
- Average running time per task.
- Ratio of CPU time used by task.
- Maximum time a task is interrupted by other tasks.

Of these five functions, the last two are particularly relevant to finding bugs or performance issues with your system.

Of course, there are many other techniques available besides the one I've focused on here. A lot will depend on the resources available in your system. For example, do you have enough CPU time and additional memory to capture and store RTOS data for later use, or perhaps, have an environment where this data can be acquired as another RTOS task and have it compressed and streamed to the host computer for analysis?

Whatever debugging tool chain solution you decide to use, you should not ignore the impact of the learning curve on your development schedule. Today, with the Internet, we are fortunate to have multiple sources of information just a few clicks away. Instead of having to send engineers to training seminars, they can watch YouTube tutorials or webinars. Ideally, the vendors have provided a robust set of training and information resources designed to teach new users how to get the most out of their complex and expensive new toys.

Case in point. I'm currently working in areas of teaching and research: Teaching microprocessor system design and studying how EE students can do their undergraduate labs via the Internet, rather than sitting in front of a solderless breadboard in the lab. For many students who need to balance school, family, and job, not having to come to campus during rush hour could be a major factor in work-life balance.

These two projects link us with two well-known companies. These companies have all kinds of online resources, videos, how-to videos, webinars, etc. But.... starting with a simple problem statement, "I need to learn how to do this (fill in the blank)," we are faced with trying to dig through many resources that are hyperlinked to many other resources. Why can't these companies invest in a software wizard? One that can analyze my query and set up the pieces of information I need in a usable format.

On the plus side, almost all tools and software packages are supported by user groups. I've found these forums to be invaluable sources of good information on how to solve gnarly problems, or just how to better use a tool. If you are really lucky, the users forum spans multiple continents, so just like the satellite communications dishes located around the world, you are never out of touch with an expert.

One of my favorite forums is the user's group that is centered around LTspice.[h] LTspice is a freeware version of the SPICE simulator that electrical engineers have been using for many years to design and simulate analog circuits. The user's forum is moderated as a mailing list.[i] You subscribe to the mailing list and, any time, day or night, you can send an email question out to cyberspace and feel pretty confident that your question will be answered within minutes from somewhere in the world. I teach LTspice in parallel with basic circuit design and the students are amazed with the good information that comes back when they are stuck on a homework problem at 2 a.m.

What all this means is that it takes time and effort to learn how to effectively use these tools. Crunch time is not the time to try to pull a tool out of the storeroom and try to figure out how to set it up, turn it on, and get useful information. That plane has left the gate. You must put time blocks in your schedule to train the engineers for the debugging requirements of the project, just as you would devote time to learn the issues of a new processor.

With an RTOS in your system, you have another set of variables to deal with. Is the problem hardware, is it software, is it in the RTOS, or is it in a combination of all of them? Furthermore, unless you wrote your own RTOS or use an open-source RTOS package, you don't know "exactly" what the RTOS is doing, so your problem-solving effectiveness will depend upon what the commercial tools can tell you when you know enough about them to ask the right question in the right way.

[h]https://www.analog.com/en/design-center/design-tools-and-calculators/ltspice-simulator.html.

[i]LTspice@groups.io.

References

[1] G. Olivadoti, S. Gollakota, Real-Time Debugging 101, https://www.embedded.com/real-time-debugging-101/, May 15, 2006.
[2] https://www.csun.edu/~jeffw/Courses/COMP598EA/Lectures/OSServices/OSServices_html/text20.html.
[3] https://www.nxp.com/files-static/training_pdf/26839_68HC08_FLASH_WBT.pdf.
[4] https://www.lauterbach.com/frames.html?rtos.html.

11

Serial communication systems

Introduction

For many years, debugging serial communications issues in embedded systems generally boiled down to debugging the vagaries of the RS-232 protocol. Surprisingly, this is still true today because RS-232C is the most basic and fundamental of the serial communications protocols and is generally rather bulletproof. Debugging typically involved getting the baud rates to properly match or messing around with transmitted data out and transmitted data in on pins 2 and 3 of the connector. Debugging the serial communications link was the first piece of I/O that the engineer needed to get working properly because communications with the target system depended upon this link working properly.

Today, serial communications protocols have greatly evolved and are being used both for peripheral communications as well as communications between elements of networks. These systems are high-speed and complex. They require highly specialized measurement tools to analyze and correct errors in the data streams. Any discussion we might have about debugging these systems would quickly focus on what company's analyzer one should purchase.

Therefore, let's narrow our scope to the types of communications systems that we would most likely have to deal with in designing real-time control systems without the necessity of resorting to specialized tools. Furthermore, we can also eliminate USB and Ethernet protocols from our discussion. You might argue that these protocols are pretty fundamental. In fact, I have a laser printer in my office that came with USB and Ethernet already installed. Shouldn't we discuss these as well?

Fair point. However, we would typically have a standard IC of some flavor that would handle the translation of the physical layer communications protocol to something that the rest of the system can then deal with. This physical layer IC circuitry is pretty standard and if you follow the design rules and circuit examples in the

Debugging Embedded and Real-Time Systems. https://doi.org/10.1016/B978-0-12-817811-9.00011-9

applications notes, your circuitry has a high probability of working correctly. However, once it leaves the translator circuit, we have to deal with it as another element of our overall system and the discussions of the previous chapters then come into play.

Here's a simple example. The early versions of the Arduino single-board computers contained a USB to UART converter IC, manufactured by Future Technology Device International (FTDI). The Atmel microcontrollers that were the core of the Arduino family of boards all had UART interfaces that could just as easily connect the chip to an RS-232 bus. The FTDI chip converts the USB protocol to UART.

Later versions of the chip, such as the ATMEGA16U2-MU, updated the communications port to directly interface to USB 2.0, eliminating the need for the FTDI interface chip. All that is now required are two 220Ω series resistors between the USB socket and the microcontroller.

So, what serial protocols should we discuss? Based upon my experiences with my student's issues with designing their microprocessor-based designs, almost all the peripheral devices they connect to their controller are either an SPI or I^2C interface. Therefore, just based upon the level of pain, let's discuss these protocols.

Because RS-232 is still around and still being used for many systems, we'll cover the basics of that protocol and the issues around getting it to work.

Lastly, and mostly because it enjoys widespread acceptance in many industries as a communications protocol, we'll consider the CAN bus. The CAN bus originally evolved as a communications standard for automotive systems, but over the years has achieved much wider acceptance in other industries as well.

Finally, the other reason that I think these four protocols make sense to discuss in a debug context is that it is possible for mere mortals to find and fix bugs using only a standard oscilloscope or logic analyzer.

RS-232

Perhaps you are familiar with the RS-232C designator. This was the version EIA[a] RS-232 standard that was in force just in the golden age of the desktop PC. The biggest change that RS-232C

[a]Acronym for the Electronic Industries Alliance. Before 1997 it was known as the Electronic Industries Association. *Source* (https://en.wikipedia.org/wiki/Electronic_Industries_Alliance).

brought was the reduction of the logic levels from ±25 V down to a more manageable ±5 V. This made it easily compatible with PCs because +5 and +12 V were the primary power supply voltages in the original PCs, along with −5 and −12 V, though at lower current levels. Later versions of the power supply standard dropped the −5 and −12 V voltages and added +3.3 V.

Without the −5 V voltage available, IC manufacturers stepped in and created interface circuits that contained their own built-in DC-DC converters. The circuit contained an internal oscillator that, when combined with external capacitors, converted +5 VDC to −5 VDC. A good example of this is the MAX232X Dual EIA-232 Driver/Receiver family from Texas Instruments.

If you were around computers and embedded systems before the introduction of USB, then you are familiar with RS-232. This is the classic COM port of the PC age. If you were around for the early days of the PC, such as the PC-XT and the PC-AT, then you remember grappling with the jumpers on the I/O boards in order to set the correct communications protocols. There were four basic sources of potential errors:

- Wrong COM port assignment.
- Improper cable pinout.
- Wrong baud rate (clock frequency).
- Improper flow control.

Wrong COM port assignment

If RS-232 is the communications protocol between your application running on your PC or workstation and your communications port on your target system, then the first order of business is to get the link working. This was crucial because this link had to be established for the host-resident debugger to be able to do its job.

So, you hook everything up, power up your target, and wait for a prompt to come back, telling you that the debug kernel in your target is talking to the debugger on the host. Nothing, nada. What's the problem? Most likely, the wrong COM port is being used. Even today, the popular Arduino IDE requires the user to select the correct COM port.

The simplest way to solve this problem is to locate any one of the free terminal emulation programs that are around. Even if you find a commercial terminal emulator, you can generally get a free trial version. With the terminal emulator installed, unplug the RS-232 cable from the target and connect pins 2 and 3 together. Type a few characters and see if they are echoed back to the screen. If so, the COM port is correct.

Now connect it back to the target and try again. If it works, great. If not, we move on.

Improper cable pinout

If you want a quick trip down the rabbit hole, this is the place to do it. The reason is that the correct cable depends upon the vagaries of the engineer who designed the mating COM port on the target system. It turns out that there are three flavors of cable connections[b]:

1. Direct DTE (computer side) to DCE (computer modem side): Here, the cable wires go straight through. Pin 1 connects to Pin 1 and down the line to Pin 9 connected to Pin 9.
2. Null-modem cable DCE to DCE (modem to modem).
3. Null-modem cable DTE to DTE (computer to computer).

Depending upon how the target system connector is wired, or which cable you happened to grab from the assortment of cables laying around, it may or may not be the right cable type. To make matters worse, there is the sex of the connectors at either end of the cable to consider.

In the ideal case, the cable should have a female connector at one end and a male connector at the other. The female end plugs into your PC and the male end should plug into the female connector on your target system. If this is the case, then there is a reasonable probability that the connection is correct. But....

That doesn't eliminate the possibility that you still have the wrong cable pinout configuration. The simplest solution is to first use an ohmmeter and determine the pinout and connectivity of the cable. If everything is connected properly as a straight-through cable, then find a null-modem adapter that does the pinout reversals that you need.

If none of these work, then it is time to bring out the oscilloscope, set it for single-shot mode, and, using the terminal program, trace the signal from the connector back to the I/O pins of your microcontroller or discrete I/O chip.[c] You should observe the serial bit stream from the character you are sending going to the received data I/O pin of the IC.

[b]https://ipc2u.com/articles/knowledge-base/the-main-differences-between-rs-232-rs-422-and-rs-485/.

[c]A discrete chip would likely be a UART, which stands for Universal Asynchronous Receiver/Transmitter. The classic UART chip was the 16550, introduced by National Semiconductor in 1987.

The key is that you now need to verify the integrity and correct operation of the I/O data transmission path. Time to employ the investigative techniques we discussed in the earlier chapters.

Wrong baud rate (clock frequency)

The tell-tale sign that the baud rate is incorrect is that the transmission is garbled.

When you press the letter A on the keyboard, you see some other character or symbol being echoed. This is a pretty straightforward problem to solve, but it is also an interesting problem to discuss.

If you look into the particulars of RS-232, you will discover that it is asynchronous (no clock). That's correct, there is no clock signal transmitted with the data. Today, the clock signals and data are encoded together and circuits at the receiving end recover the clock in order to synchronize the transmission.

With RS-232, the receiver and the transmitter each have their own clock. The clocks should be running at the same frequency, but they are not synchronized to each other. In other words, the phase relationship between the two clocks is entirely random. So, this issue becomes how to accurately transmit and receive the data if each system's clock is not synchronized with the other? This is the clever solution.

First, without getting lost in the arcana of communications protocols, we can simply say that the baud rate (9600 baud, for example) is the bit rate for data transmission. If every data transmission contains 1 start bit, 8 data bits, and 1 stop bit, then transmitting something requires that we send 10 bits of data, or 9600 divided be 10 characters per second. Therefore, 9600 baud would translate to about 960 characters per second. At 9600 bits per second, we need to clock the data in every 104 μs, which requires a clock frequency of 9.6 kHz.

Here's where it gets interesting. We still need to solve the problem of the unsynchronized clocks. This occurs at the receiving end. The 9.6 kHz data clock is actually divided down by a factor of 16 × from the master clock. When the receiver sees the negative edge of the start bit, it counts 8 cycles of the master clock and then begins to clock in the data every 16 clock cycles of the master clock.

By counting eight cycles, it starts to clock the data roughly in the middle of each bit being transmitted. The uncertainty in the phase relationships becomes negligible because the phase difference is in the master clock, not the data clock. As far as the data is concerned, the correct clock edge to bring in the data occurs more or less in the middle of the data bit.

This is also the reason why the maximum baud rates for RS-232 top out at around 56K baud. As the data rate goes up, the phase uncertainty becomes more significant. Also, RS-232 cables are not impedance-controlled transmission lines and the longer the cable and the higher the data rate, the more distorted the transmitted data become. With lower margins of error, bit errors are more likely.

From my perspective as a teacher, the operation of a UART is a great exercise to give students learning about finite state machines and simple UARTs can be constructed in their FPGA lab experiments. Anyway, that's how the baud rate system works and the necessity to synchronize the baud rates comes about because the transmitter and receiver clocks are not phase locked to each other.

Improper flow control

If everything else is working, the last possible data transmission error source is flow control. There are three possibilities here:

1. No flow control.
2. Hardware flow control.
3. Software flow control.

If both systems are set up for no flow control, then it is assumed that the receiver is fast enough to accept as much data as the transmitter can send without worrying about any input buffers being overflowed. Once a transmission starts, no matter how much data is being sent, the receiver can deal with it.

I think you can see what might happen with this type of communications channel and an RTOS on the target system. If the priority level of the communications channel is too low, and there is no flow control in place, then data can be lost.

With hardware flow control, the two signals, clear to send (CTS) and request to send (RTS), provide the handshake mechanism between the receiver and transmitter. The transmitter asserts RTS in order to initiate a transmission and the receiver responds with a CTS handshake. This informs both devices that data can now be sent. If the receiver cannot keep up with the transmitter, it deasserts CTS and the transmission stops until both devices are ready again. For this to work, these pins must be activated in the cables, the connectors, and in the UART driver code.

It was fairly common to save space and simplify the connection by using a simple telephone jack for the connector. In this case, only ground, send, and receive were used. With this situation, flow control (if used at all) became a software issue. Two ASCII control codes were used for flow control, XON (hexadecimal 11) and XOFF (hexadecimal 13).

The data transmission handshake now depended upon sending and receiving these control characters to properly pace the data flow. Of course, the software drivers at both ends had to agree on the type of protocol being used for this system to work.

I2C and SMBus protocols

Peripheral chips, such as A/D and D/A converters, timers, and UARTs, used to all be parallel interfaces with the processor. You connected them to the address and data buses, used some sort of address decoder scheme to assign them a memory or I/O address, and that was it. Assuming your timing margins of your design met the specifications of the peripheral device, the device should work as advertised.

As microcontrollers became more integrated, external buses disappeared and a different method was needed to interface peripheral chips to the controller. Enter the onboard serial buses, I2C and SPI. Today, peripheral chips with parallel interfaces are becoming endangered species. Out of curiosity, I went to the Analog Devices website[d] and looked at the product selector guide for one type of A/D converter. There were 338 total parts listed for single-channel A/D converters and less than one-fourth of them contained parallel interfaces. Most of those were combination parallel and SPI devices. Less than 20 were pure parallel interfaces.

I2C was developed by Philips and its key feature is its simplicity. It is interchangeably described as either I^2C or I2C. The actual name is "interintegrated circuit" bus. Surprisingly, it is also the basis of the CAN bus that we'll discuss later in this chapter.

The SMBus (system management bus) was developed by Intel and Duracell in the mid-1990s as a simple two-wire bus for use in smart batteries and on PC motherboards.[e] It is hardware compatible with I2C with some differences [1, 2]. The major difference is that the SMBus is a low-speed bus, limited to a maximum clock frequency of 100 kHz. There are other differences having to do with logic voltage levels and time-outs, but for our purposes they may be considered to be compatible because I2C devices may be used in SMBus applications with the caveat that differences exist.

[d] https://www.analog.com/en/parametricsearch/11007.
[e] http://smbus.org/.

However, because I2C dominates in the embedded realm, we'll continue to focus our attention there.

As previously mentioned, the I2C bus is a two-wire bus, and because it is an open collector (open drain) topology, it requires pull-up resistors at the end of the run. However, with open-collector (or open-drain) outputs, multiple drivers and receivers can be attached to the same wires.

Because the outputs are not actively driven, the low-to-high transitions are controlled by the RC time constant comprised of the pull-up resistors and the total capacitance of the wires and I/O devices connected to those wires. Thus, I2C bus speeds are restricted to data rates below 3.4 Mbps (megabits per second).

I2C uses the concept of a master/slave relationship between the devices. All data transfers are initiated by the master and the master provides the clock signal, serial clock (SCL). Data transfers are bidirectional over the serial data line (SDA). Fig. 11.1 shows this schematically. Note that only one master device is shown but the standard allows for multiple masters as well as multiple slaves (as shown).

The most common source of error that I've observed with my students is the omission of the pull-up resistors. Without the pull-ups, we're looking at one end of a transistor flapping in the breeze.

Fig. 11.2 shows the I2C data transfer protocol. As previously mentioned, a data transfer must be initiated by a master device by bringing SDA low while SCL is low. Data on SDA may change while SCL is low but must be stable while SCL is high.

As shown in Fig. 11.2, data is transmitted MSB first in 8-bit long data packets, although as many packets as necessary can be exchanged between devices. After each packet is sent, the receiving device will send an acknowledge signal back so that the next packet may then be transmitted.

Addresses are assigned for each device by the I2C bus committee. So, you might ask how you can put several identical slave

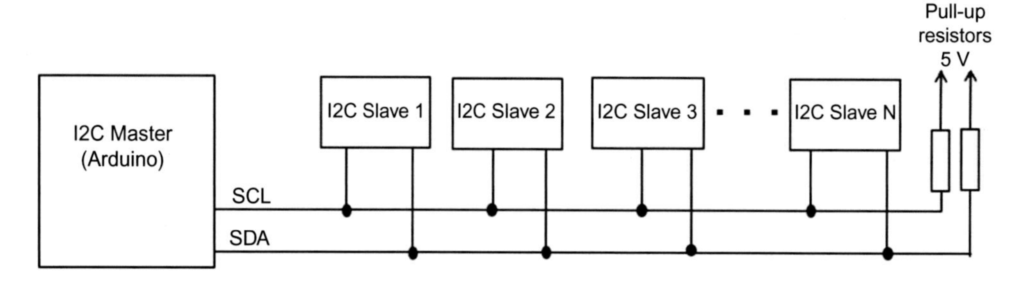

Fig. 11.1 I2C interface diagram. Data transfer begins by bringing the SDA line low while the SCL line is high.

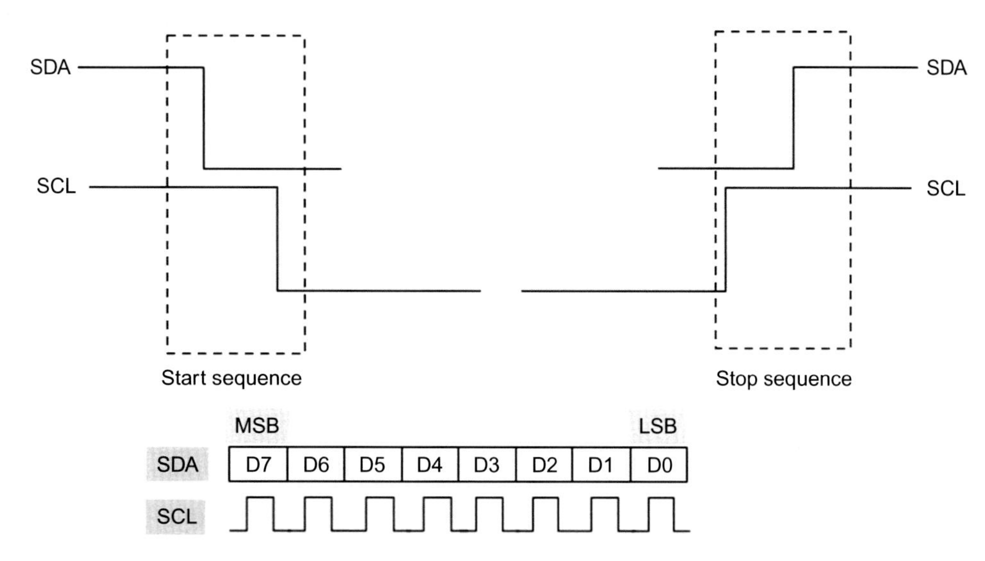

Fig. 11.2 I2C data transfer protocol.

devices on the bus without conflicts? Even though each device has a unique 7-bit preassigned address, it is possible to partition this address into a smaller fixed part and a programmable part that depends upon a nonvolatile memory entry or programmable pins on the device. Thus, the 7-bit address can consist of 4 fixed bits and the state of 3 I/O pins tied to ground or power (Vcc). Some devices with nonvolatile storage, either EEPROM or FLASH registers, can have their I/O address reassigned by the customer under program control.

One source of possible error is addressing slave devices. When sending out an address, the master sends out the 7-bit slave address, which is left-justified toward the MSB, and the eighth bit (LSB) indicates whether the master is going to write to the device (LSB = 0) or read from the device (LSB = 1). From the point of view of programming the address, the byte value containing the address must be shifted left one bit while making sure that the instruction pads the LSB with a zero bit. Next, a byte containing all zeros plus the read or write information in the LSB position is OR'ed. with the address to complete the addressing code. Thus, if the device address is 04H, and the master is writing to the slave, the byte value sent out would be 08H.

The I2C specification also allows for 10-bit addresses [3]. A special 7-bit address consisting of the sequence 1 1 1 1 0 X X alerts slave units with 10-bit addresses. The X X bits of the last part of the address are not "don't care bits." They are the first 2 bits of the 10-bit address. The next 8 bits of the following data packet complete the 10-bit address.

I was first introduced to the potential issues surrounding I2C addressing modes when one of my students was involved in a Capstone project with a company located in our area. I'm not going to mention the company nor the nature of the project, other to say that this Capstone team of four students was developing a new microcontroller board for the company. The company's engineering mentor to our team gave us the board they wanted to modify and all the software, which, we were told, "worked fine."

One of the students was working on communications with the peripheral devices, which used a I2C bus. Everything seemed to be working except he could not talk to one I/O device on the board. Everything he tried failed. In desperation, after a week or so, he asked for my help. The first thing I suggested is to put a scope on the SCL and SDA lines and look at the signals going out to the device.

We did that and everything looked fine as far as the software was concerned. The 7-bit address matched the "C" code. Next, we grabbed the data sheet for the part and started to read. Under the addressing section, we noted that this device had a 10-bit address. Either we had the wrong software, or it never worked in the first place. After fixing the code, it worked fine.

Reference [3] gives a really nice overview of the protocol and all the minutia that would cause our eyes to glaze over, but if you have a bug that you are trying to track down, this is a very readable document

From the point of view of debugging, I2C is a relatively easy protocol to debug. Because it has only two wires, it is easy to clip the probes from a two-channel oscilloscope onto the SDA and SCL lines and watch as the traffic goes by.[f] If you happen to have a logic analyzer handy, many of them also support I2C bus protocol analysis through postprocessing of the data. The LogicPort logic analyzer (previously discussed) is one such device. Clip two LA channels onto your SDA and SCL lines, set the interface, and you have a I2C (or SPI) data analyzer.

There are also reasonably priced, dedicated I2C tools available from a number of companies. Of particular note is the BusPro-I bus analyzer, monitor, debugger, and programmer from Corelis.[g] I'm mentioning Corelis here because I'm quite familiar with the company from my days working on development tools at

[f]Of course, you had the foresight to place test points on these signal traces just for this purpose.

[g]https://www.corelis.com.

Advanced Micro Devices (AMD) working with third-party vendors to provide design and debug tools for the Am29000 family of embedded microprocessors and microcontrollers.

Corelis worked closely with AMD to provide JTAG support for the on-chip debug capability and boundary scan. Finally, I have no financial relationship with the company and can't comment on their products versus the competition. Simply, they've been around for a long time and make tools that do the job. From our perspective, the BusPro-I offers two very significant debugging capabilities:

1. Monitoring, recording, message filtering, symbolic translation, and conformance to protocol.
2. Interactive debugging with the ability to drive data to the I2C bus to provide stimulation and response to devices on the bus.

Fig. 11.3 is a screenshot of the BusPro-I monitor window. Another attractive and useful feature of this tool is the flexibility of the user interface. The software is also available as a C/C++ library of functions that may be integrated with other debugging tools.

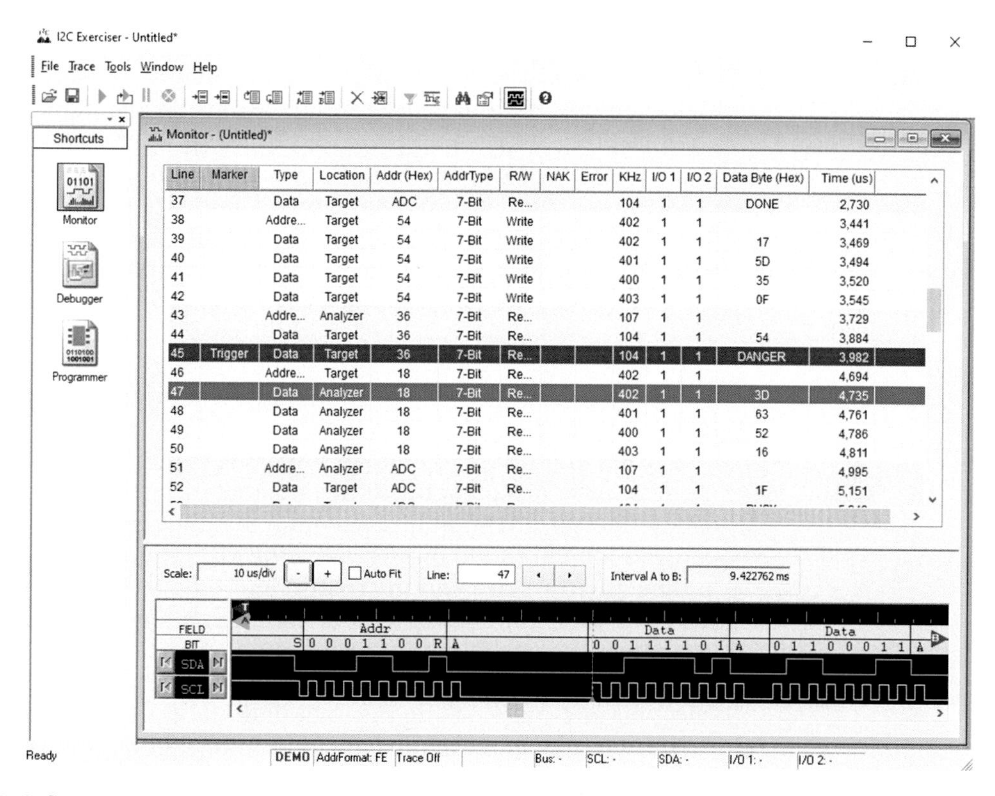

Fig. 11.3 Screenshot of the Corelis BusPro-I I2C bus analyzer, monitor, debugger, and programmer. Courtesy of Corelis.

The BusPro-I is more costly than the other I2C tools that are available, but it is a professional-grade tool from a company that's been in business for quite a while. If you are looking for a tool, I might suggest that you use this tool as your reference and measure the others against it.

On the other hand, it is a specialized tool, capable of debugging significant problems in I2C systems. Is it worth the incremental cost of using an oscilloscope or a logic analyzer? You'll have to judge that for yourself. A for-profit company understands the time value of money. Saving several hours of debugging by having the right tool for the job could easily justify the cost of the tool. For the student, a lab scope or logic analyzer works just fine and is free.

SPI protocol

The serial peripheral interface (SPI) was invented by Motorola in the mid-1980s and along with I2C, it has become a *de facto* standard for short-range, high-speed, interdevice communications. However, it has not been codified by any standards organization or industry-based standards group. If you read some of the embedded systems forums where embedded engineers share their opinions, SPI is an easier interface to code and debug.

Unlike I2C, SPI is a point-to-point protocol. There is no bus in the same sense as I2C, but SPI devices can be configured in much the same way that we build memory systems. That is, each slave device has its own active-low, chip select input. The master device must uniquely activate the slave it wants to communicate with. On first glance, this might apparently complicate the hardware, but given that microcontrollers typically have a rich set of parallel I/O bits to use, this is less of an issue. The advantage is that it is not necessary to deal with transmitting address information to the peripheral devices.

The SPI interface is full duplex, meaning that the slave is sending data to the master at the same time that the master is sending data to the slave. The implementation is rather clever. The data path is a loop from the master along the MOSI (master out, slave in) line into the slave and the return line is the MISO (master in, slave out) line. The master and slave each contain a shift register that clocks the data out and in at the same time.

Sending data out to a slave involves setting the correct chip select bit low and clocking the data until the desired number of data bits has been sent. Thus, there is no restriction that all data packets are byte-wise, as is required by I2C. Also, there is no need for the slave to send an acknowledge signal at the end of the data transfer.

As defined, the SPI interface is a three-wire or four-wire "almost" synchronous point-to-point system. The interface is not fully synchronous because the chip select signal may be asserted at any time, rather than synchronized with the clock. The normal arrangement is four-wire full-duplex, but in certain situations, such as a D/A converter, only MOSI mode is appropriate, so three wires are all that is necessary. Because the master and slaves (no multiple masters are allowed) are typically resident on the same PC board, there is no need to have a power or ground signal between them. When used in three-wire mode (half-duplex), the protocol becomes more like I2C because only one device can be sending at a time.

The clock is controlled by the master and both phases are used. There are four modes of setting up the clock phase and data transmission relationship, not surprisingly named Mode 1 through Mode 4. The most appropriate operational mode for a given situation is left as an exercise for the designer, but according to the Analog Devices application note [4], the master should be able to support all four modes, although this is not a requirement for the slave devices.

Two phases of the clock are necessary because the same clock is being used to transmit and receive the data. Thus, if the data are sent on the rising edge of the clock, they can't be read on the same rising edge. Therefore, if we send the data on the rising edge, we read them on the falling edge, giving the receiver sufficient time for the data to be stable before clocking it into the receiver's shift register. If multiple slaves each are designed to use a different mode, the master will have to reconfigure its configuration register to make it compatible with the slave.

One of the most common areas for trouble centers around the four modes of clocking and data transmission. According to Williams [5], potential problems with phase and polarity are the first places to look if you need to troubleshoot your SPI protocol. Here's how Williams explains it,

> *The choice of which edge to read data on, as well as whether the clock signal idles high or low, presents two binary variables that can change from one chip to the next, giving us four different "versions" of SPI. The idle state of the clock signal is called clock* polarity, *and it's easy to explain. A clock that idles high has a polarity = 1, and vice-versa.*
>
> *Unfortunately, if you like thinking about when in the clock cycle the chip reads the data, the industry decided to latch on to another aspect of the transmission which maps to the same thing: the* phase. *Phase describes whether the data is going to be read on the first clock transition (phase = 0) or the second (phase = 1). If the clock idles*

low (polarity = 0) the first transition is going to be upward, so a system that samples on the upswing will have phase = 0. If the clock idled low, however, the first transition is necessarily down, so a system that samples on the upswing will have phase = 1 — sampling on the second transition. My head hurts even writing it out.

Here's how I cope. First, I look at when the data is sampled. If data is sampled on the upwards clock edge, the phase equals the polarity, otherwise it's the opposite. A read-on-rising-edge is 0,0 or 1,1. And since the polarity makes sense, it's easy to pick between the two. If it idles low, you have 0,0.

	Sample on rising edge	**Sample on falling edge**
Clock idles low	Phase: 0 Polarity: 0	Phase: 1 Polarity: 0
Clock idles high	Phase: 1 Polarity: 1	Phase: 0 Polarity: 1

Williams goes on to discuss each potential problem area. Rather than do a paste and copy of his original article, I'll just touch on the high points. Other potential areas for problems can be summarized below:

- Speed: Because the master sets the clock speed for both devices, a simple approach is to slow down the clock. If the slave device can't keep up with the master, then problems may arise. The master may have a hefty MOSI driver on the output while the slave does not. A two-channel scope with a separate trigger input from the clock can give you some insight into the error margins of the circuit. If the master clock is 10 MHz, slow it down to 1 MHz, or as Williams suggests, 100 kHz.

  Once you have it running reliably, gradually raise the clock frequency until you begin to see errors. However, keep in mind that the problem may not be simply clock speed, but rather issues related to the set-up and hold times [4] due to the mode you're using.
- Insufficient clocks: If you are expecting to read a slave and the data never come, it may be due to the fact that the master has stopped issuing clock pulses. Recall that the master must continue to supply clock pulses until the slave has finished sending all the data, and data packets are variable length. Therefore, if this is your issue, start looking at your driver code and make sure there is a clear handshake as to how much data will be transferred, or if you have a termination string at the end of the packet.

- Bus problems: When a system goes beyond one master and one slave, then problems traceable to bus contention issues can crop up. This can be as simple as neglecting to make sure that only one chip select output from the master is low at a given time.
- Open collectors: Some slave devices use open collector outputs to drive the MISO line. They may have a weak pull-up resistor or none at all. Unless you catch this in the data sheet, you may have neglected to add a pull-up resistor to the MISO line. Williams suggests that a good way to test for MISO problems is to bias the line at the midpoint of the logic swing using two large resistors, 100 kΩ, for example. Set the resistors as a voltage divider between power and ground with the MISO line biased at the midpoint. With all chips turned off (Chip Select = 1), see if the line is pulled up to Vcc or pulled down to ground. If either one of these situations is true, one of the slaves is not turning off.

  Next, run the system and observe the bus. If you see what looks like valid data, but the logic swing is from the midpoint to ground instead of power to ground, you have an open collector issue.
- Best practice: During initialization of the microcontroller, the SPI devices can be subject to various inputs, floating chip select lines, and random noise. Placing weak pull-up resistors on the chip select lines guarantees the slave units will remain in the off state until the microcontroller is fully awake and in control.

Tools

Fortunately, SPI is such a popular protocol that there is a wealth of information available on how to debug SPI devices and the tools available to effectively troubleshoot any problems. Corelis offers the BusPro-S, which is a full-featured SPI host, debugger, and programmer with capabilities similar to those of their previously mentioned I2C unit. In a sales-biased application note,[h] Keysight [6] discusses troubleshooting today's most common serial buses; I2C, SPI, USB, and PCI-Express Generation 1.

Keysight's Infiniium oscilloscopes contain protocol analyzers as well as the traditional oscilloscope functionality. Because these instruments are for the most part software-based, I'd be reasonably confident that as new communications protocols emerge, software upgrades would be available.

[h] This isn't necessarily a reason to ignore it. Most application notes are geared to encouraging you to use their product.

In a white paper on the subject, Leens [7] nicely summarizes the various protocols and different techniques of testing and debugging them. In particular, he makes the point (which never occurred to me, so I found it really interesting) that chips such as FPGAs and CPLDs can be debugged using the SPI interface to talk to their JTAG port. Why? Because JTAG and SPI are both clocked loops. You can read all the internal JTAG registers using the SPI interface. The downside is that JTAG loops can be very, very long, making for rather low speed communications.

Controller area network (CAN-bus)

If you drive an automobile built from the 1990s on, then there is a high probability that the car you're driving is using a CAN bus for interprocessor communications because the CAN bus was originally designed to be an automotive-based bus. The CAN bus was originally developed in 1983 by the German company Robert Bosch GMBH, and officially released in 1986. The first controller chips, manufactured by Intel and Philips, were released in 1987.

Since its introduction as an automotive standard, the applications of the CAN bus have expanded to industrial applications, medical, and military as well as other applications that face harsh physical environments [8].

For signs of how widely CAN bus applications have grown, you can find CAN bus cards (Arduino Shields) for most of the mainstream single-board computers. The ISO11898-2 and ISO11898-5 specifications provide details for the high-speed CAN physical layer or transceiver as described by Monroe [9]. According to the author, finding common problems with CAN bus operation is relatively straightforward and may be accomplished with the most basic debugging tools. Of course, like any other widely accepted standard, there are specialized tools available at various price points with all levels of capabilities.

The LogicPort logic analyzer that I've discussed in this and previous chapters also includes a CAN bus interpreter as part of its suite of standard bus protocol interpreters. Probably the most telling example of the wide acceptance of the CAN bus is an advertisement for a CAN bus analyzer for $150 from Microchip being sold through Walmart. Microchip also has a YouTube video[i] on the use of the 63r7680 Microchip APGDT002 can bus analyzer tool.

[i]https://www.youtube.com/watch?v=WfSZdWHiM9k.

Returning to the Monroe article, the author states that with an ordinary digital multimeter, a power supply, and an oscilloscope, most of the most common CAN bus problems can be identified and repaired.

At this point, I am going to close this chapter of the book. Honestly, I'm not trying to shortchange you, gentle reader. The two articles that I've cited here, the EEHERALD article and the Monroe article, are excellent sources for understanding the protocol of the CAN bus (EEHERALD) and how to debug the most common CAN bus problems (Monroe).

Closing remarks

I enjoyed writing this chapter because I learned a lot doing the research of the material for it. I have lots of experience debugging RS-232C and its higher-speed variants; many of my students debug problems with I2C and SPI peripheral devices on their project boards. Also, because my background is designing and debugging microprocessor systems, I'm a lot more comfortable plugging a logic analyzer onto address, data, and status buses and observing 100+ logic signals in parallel than I am trying to make headway swimming against a serial bit stream.

If I were to take the time and condense the Monroe and EEHERALD articles, I think we would both get tired constantly seeing my citations because I don't plagiarize and I'm merciless with students who do. So, if you are having CAN bus problems, you won't go wrong taking an hour and reading these two articles.

Additional resources

1. Editorial Staff, Ease the Debugging of Serial Peripheral Interfaces, Electronic Design, https://www.electronicdesign.com/technologies/boards/article/21767094/ease-the-debugging-of-serial-peripheral-interfaces, September, 2001.

References

[1] Texas Instruments, *SMBus Compatibility With an I^2C Device*, SLOA132, April, http://www.ti.com/lit/an/sloa132/sloa132.pdf, 2009.
[2] Maxim Integrated Products, Comparing the I^2C Bus to the SMBus, Applications Note 476, https://pdfserv.maximintegrated.com/en/an/AN476.pdf, December, 2000.
[3] https://i2c.info/i2c-bus-specification.
[4] M. Usach, SPI Interface, Application Note AN-1248, Analog Devices, Norwood, MA, 2015.

[5] E. Williams, What Could Go Wrong: SPI, https://hackaday.com/2016/07/01/what-could-go-wrong-spi, 2016.
[6] Keysight Technologies, Strategies for Debugging Serial Bus Systems with Ininiium Oscilloscopes, Application Note 5990-4093EN, July, 2014.
[7] F. Leens, Solutions for SPI protocol testing and debugging in embedded system, White Paper, Byte Paradigm, www.byteparadigm.com, August, 2008. https://www.saelig.com/supplier/byteparadigm/BP_UsingSPIForDebug_WP.pdf.
[8] EEHERALD Editorial Staff, Online Course on Embedded Systems—Module 9 (CAN Interface), http://www.eeherald.com/section/design-guide/esmod9.html, December, 2016.
[9] S. Monroe, Basics of Debugging the Controller Area Network (CAN) Physical Layer, Texas Instruments Applications Note, http://www.ti.com/lit/an/slyt529/slyt529.pdf, 2013.

12

Memory systems

Introduction

You could say that we've saved the best for last. Memory is the heart of your system. As with most aspects of embedded systems, memory debugging has a hardware and a software component. Before we can debug software issues, we need to make certain that the hardware is both functional and reliable. Therefore, I would argue that the first testing and possible debugging that should be done must be the processor-to-memory interface.

If the memory is internal to your microcontroller, there isn't much debugging to do outside making certain that all the configuration registers are properly initialized. If the microcontroller is also addressing external memory, then the same issues arise as with any external processor/memory interface.

In this chapter, we'll first explore the basics of memory debugging using static RAM (SRAM) as our model and then move on to a general discussion of DRAMs and DRAM debugging methods. Discussing the nature of how RAM works is crucial to having insights into the possible root causes of an error or defect.

After discussing hardware-based error-chasing techniques, we'll look at memory bugs that are software-based.

General testing strategies

In a very thorough article on testing RAM in embedded systems, Ganssle [1] discusses general strategies for testing RAM memories in embedded systems. His first observation is that the RAM test should have a purpose. He says,

> *So, my first belief about diagnostics in general, and RAM tests in particular, is to clearly define your goals. Why run the test? What will the result be? Who will be the unlucky recipient of the bad news in the event an error is found, and what do you expect that person to do?*

Debugging Embedded and Real-Time Systems. https://doi.org/10.1016/B978-0-12-817811-9.00012-0

If you suspect a hard RAM failure, it may manifest itself in several ways. If you're lucky, the system will detect it, report it, and try to recover. If not, the system will crash. If this is an intermittent problem, you may not learn about it until customers start to complain, or there's a catastrophic failure and people are injured.

Ganssle goes on to debunk my favorite memory test pattern, alternating 0xAA and 0x55 bit patterns, and provides good reasons why these patterns are rather poor for any kind of extensive and thorough memory test. One reason that the alternating 0xAA/0x55 test is so poor is that if the problem has to do with memory addressing, there is a 50% probability that writing and reading back from the wrong address will still pass as good memory.

He suggests constructing long strings of almost random bit patterns and testing with them. If the string happens to be a prime number (Ganssle suggests 257), then repeatedly writing and reading to blocks of memory will not map in multiples of memory blocks with the same address bits, differing only in the higher order bits.

Another good tip from his article is to write an entire block first, then read it back and compare. Testing small blocks at a time will not find all errors.

Poor memory design will often manifest itself as sensitive to a particular bit pattern while working perfectly with most other random reads and writes. While we might attribute this to a bad chip, the likelihood, as Ganssle points out, is more likely to be bad PC board design practices, such as inadequate power bus filtering, electrical noise, inadequate drive capabilities, or one that can sneak up on a digital designer, the dreaded "analog effects."

Most digital designers, including me, would almost never think about terminating address, data, and status buses between the processor and the external memory. A good design guideline [2, 3] gives a general rule of thumb for deciding when you need to properly terminate a trace in its characteristic impedance.

The speed of a signal on a printed circuit board is approximately 6–8in./ns, about half the speed of light in free space. Assuming the value of 6in./ns, then you should terminate the trace if the trace length is greater than 3in./ns of rise time of the trace. So, a pulse with a rise time of 1ns can be left unterminated if it is less than 3″ in length. Of course, as edges get even faster, this value will decrease even more. This is pretty sobering if you are working with gallium-arsenide (GaAs) logic with rise times around 100ps. There, any trace over approximately 5/16 to 3/8 of an inch needs to be terminated. Fortunately for most of us, that's a different world.

When and if to choose to use line driver circuits can be a tough decision. You'll incur power, access time, board area, and cost penalties if you use drivers. However, memory arrays, or particularly DRAM arrays, as Ganssle points out present a large capacitive load to the output pins of the processor writing to memory. Most microdevice outputs are not designed to drive capacitive loads. As a result, rise times may become unacceptably slow, leading to noise and instability problems.

Static RAM

Whether you are using a microcontroller with onboard RAM or a microprocessor with external RAM, the workings of the system in both cases are the same. As its name implies, this RAM is static. As long as power is applied, the data stored within its cells are stable. Writing to the RAM cell can change the data and reading from the RAM cell leaves the data unchanged.

Static RAM can be very fast and is always accessible. On the other hand, dynamic RAM contains overhead that will sometimes prevent immediate access to it. Why then isn't static RAM the dominant memory architecture in modern computers? The simple answer is density. Dynamic RAM requires one transistor to store one bit of data. A static RAM cell typically requires four or six transistors to store one bit of data.[a]

Fig. 12.1 is a schematic representation of a typical static RAM cell consisting of two inverters, A and B, and two switches, C and D. Each inverter contains a CMOS pair and each switch is a single NMOS transistor, thus yielding a six-transistor memory cell.

Referring to Fig. 12.1. Assuming that the value at switch D represents the data stored in the cell, if switch C is momentarily connected to ground, the output will flip from a 0 to a 1 and switch C can be opened without changing the data stored in the cell.

SRAMs excel in any application that requires completely random access to any address in memory with no latency. DRAM, though much denser, is superior in applications where sequential access is the most common because the DRAM architecture is optimized for operating with on-chip caches in the CPU. As a simple figure of merit, the largest SRAMs are of the order of 16–18 megabits (Mbits) while DRAMs can be found up to 16 gigabits (Gbits). More on this later.

[a]Introduction to Cypress SRAMs, Application Note, AN116, Cypress Semiconductor Corporation, October 2006.

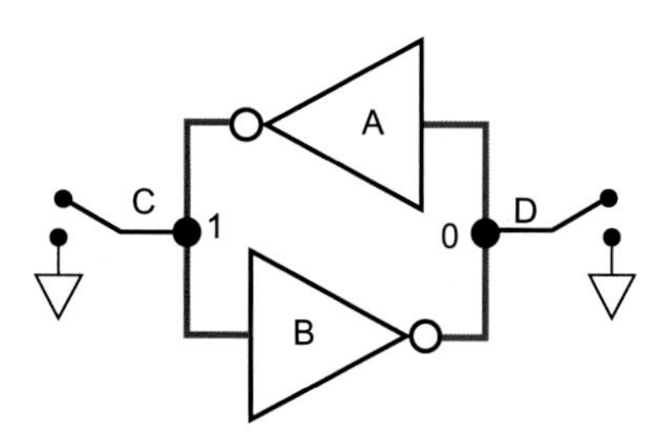

Fig. 12.1 Schematic representation of a static RAM cell. Inverters A and B each contain a CMOS transistor pair.

SRAMs come in two main varieties, synchronous and asynchronous. Asynchronous is the easier to understand, so we'll consider that one first. Fig. 12.2 is a greatly simplified schematic diagram of a typical 32 Kb (256 Kbit) SRAM.

The SRAM has 15 address inputs, labeled A0–A14, and 8 inputs and outputs that would be connected to the processor's data bus.

There are three control inputs, all active low. These are:

- Chip enable $\overline{CE}$: Master chip controller. Must be asserted for any read or write operation.
- Write enable $\overline{WE}$: Must be asserted when data is to be written into the device.
- Output enable $\overline{OE}$: Must be asserted when reading the contents of the memory is required.

Note that $\overline{WE}$ and $\overline{OE}$ are mutually exclusive. You should not assert them both at the same time.

The operation of the asynchronous SRAM circuit is quite straightforward. Let's consider a READ operation. The 14 address bits are applied to the device and must be stable for some period of time before the chip enable or output enable input are asserted. With the address bits and control inputs stable, data will appear of the 8 I/O pins and may be read by the processor after the appropriate time delay, called the access time, occurs. Fast SRAMs have access times of 15 ns or faster while slow SRAMs might have access times 40 ns or higher. Anything between the two is up to the marketing department.

On the other side of the wall, we have the processor. It has its own set of timing specifications dictated by its clock and finite state machine that controls bus operation. Suppose our memory has a 50 MHz clock and every memory cycle requires 4 clock cycles. For simplicity, we'll further subdivide these 4 clock cycles into 8 half clock cycles and label them φ1 through φ8, representing eight phases of the state machine.

Referring to Fig. 12.3, a memory cycle starts with a new memory cycle starting in φ1. Addresses begin to change and are stable

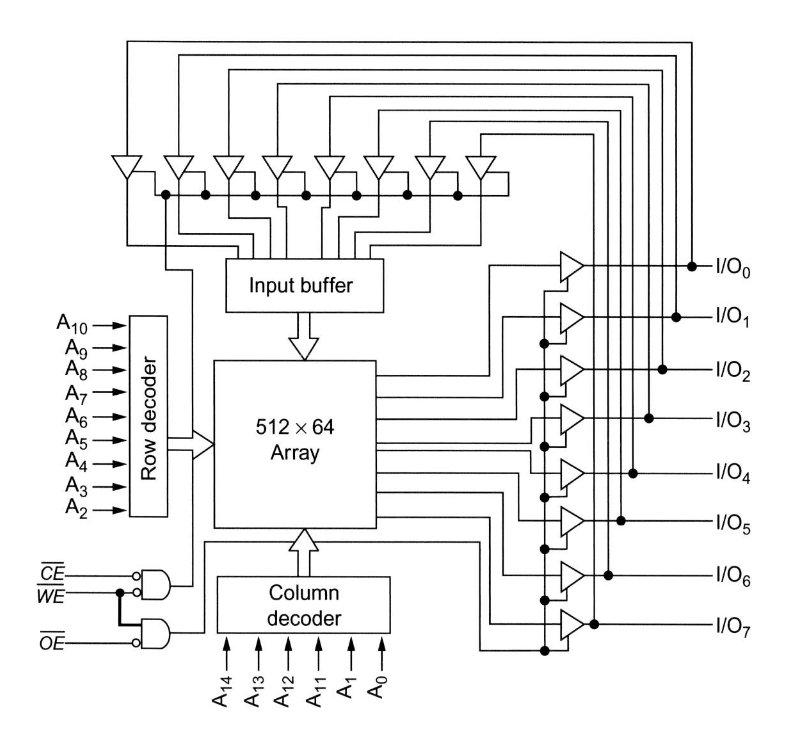

Fig. 12.2 256 Kbit SRAM organized as 32 Kb deep by 8 bits wide.

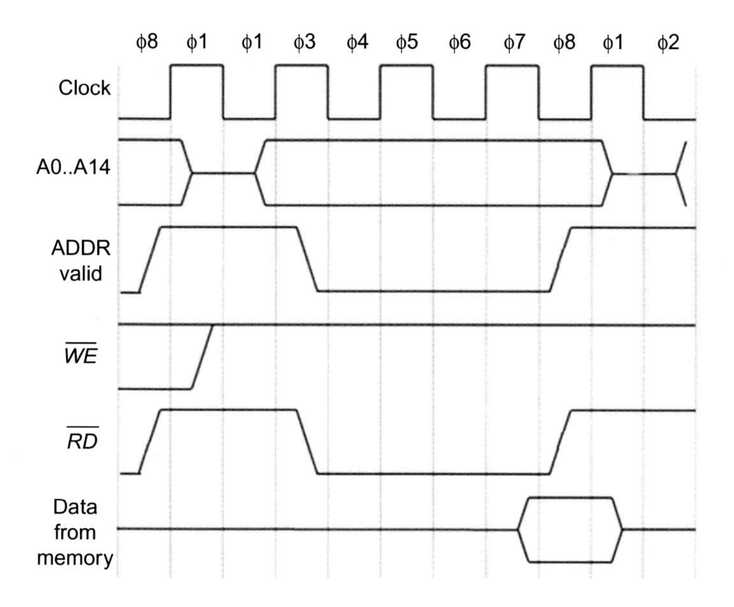

Fig. 12.3 Timing diagram for a representative 8-bit processor. φ1 is the start of the memory READ cycle and φ8 is the end of the cycle. Additional cycles, φ1 on the left and φ1 and φ2 on the right, show the end of the previous cycle and the beginning of the next cycles on the right.

around the beginning of φ3. This is indicated by the assertion of the address valid (ADDR valid) signal about midway through φ3. At the same time, the read ($\overline{RD}$) signal becomes active and, as they say at NASA, "The clock has started."

We can see from the diagram that the READ signal starts halfway through φ3 and ends halfway through φ8. Therefore, we have a total of 5 half clock cycles. Because we have a 50 MHz clock, each full clock cycle is 20 ns. Thus, we have 5 half clock cycles and each half clock cycle is 10 ns, giving a total time duration of 50 ns from the time that the processor asserts that a memory READ operation is taking place to the time that it actually reads in the data from memory on the rising edge of the $\overline{RD}$ signal.

Any static RAM memory faster than 50 ns will work in this application. Any memory with a rated access time of 50 ns or slower *may work* in this application, but the manufacturer will not guarantee it under all operating conditions.

What I haven't shown on this simplified timing diagram is how the processor deals with memory that may have access times longer than 50 ns. We could slow down the processor's clock, but that slows everything down. The best solution if we can't get faster memory devices is to extend the time duration from the assertion of READ to the deassertion of READ. There are various hardware methods to do this, but they all fall under the general umbrella called "wait states." For example, a wait state in this system could simply extend the φ5 and φ6 phases one full clock cycle, giving us φ5, φ6, φ5′, and φ6′. Adding this wait state changes the maximum access time requirement from 50 to 70 ns. Adding more wait states just continues the process until the memory access time meets our specification.

Some processors have programmable wait state registers that can be used for specific memory regions. ROM memory is typically slower than RAM memory, so if you are seeing intermittent failures, one debugging strategy would be to add a wait state and see if that improves it.

As an example, the NXP ColdFire microcontroller family was a very popular successor to the original Motorola 68 K family. It was "mostly" code compatible, with some 68 K instructions no longer supported. Of note for this discussion are the eight internal chip-select registers that allowed the user to program:

- Port size (8-, 16- or 32-bit).
- Number of internal wait states (0–15).
- Enable assertion of internal transfer acknowledge.
- Enable assertion of transfer acknowledge for external master transfers.
- Enable burst transfers.

- Programmable address set-up times and address hold times.
- Enable read and/or write transfers.

The chip-select registers could each be independently programmed for the type of memory and its access characteristics occupying up to eight different regions of memory.

I have some familiarity with this processor because I used it when I taught an embedded systems class for computer science students, rather than electrical engineering students. In their lab class, they had to learn to wade through the dense sea of acronyms in the user's manual to try to figure out how to correctly program the registers with their set-up code before the processor is ready to accept application code.

In the context of debugging, properly programming the many internal registers of a modern microcontroller is not a task for the weak of heart. Microcontrollers should come out of RESET in an operational limp-mode. This enables the programmer to be able to set up the operational registers in order to establish the runtime environment. Making sure that these registers are correctly set would be the first step in any memory debugging plan.

Here's one scenario. Everything seems to be working but the system seems to be running slower than it should be. Rechecking the chip-select registers shows that you've programmed four additional wait states by placing a 1 bit in the wrong field of the register.

When we have a microcontroller with internal chip-select registers, then the timing diagrams for external memory accesses will be correct as presented because the chip manufacturer has already accounted for the time to generate the chip-select signals in the timing specifications. However, if you are designing the external memory system and you are also designing the decoding logic, then you must also allow for the decoding logic for the external memory.

Here's a simple example. Let's assume that you have an 8-bit processor accessing external memory. The external memory is comprised of four static RAMs. Each SRAM is organized as 1 Meg deep by 8 bits wide. Your processor is capable of addressing 16 Meg of external memory, requiring 24 external address lines. Each SRAM chip requires 20 address lines, or A0–A19, 2 more address lines are used to select 1 of the 4 SRAM devices, A20 and A21, and 2 address lines, A22 and A23, are not needed.

You could leave A22 and A23 unconnected but then you might have to deal with the SRAM appearing 4 times in the processor's external memory space because the decoding logic won't decode those 2 address bits. I would include them in my logical equations, but that is your call.

There are many different ways to implement the decoder, and my intent is not to go through a design exercise. Let's just say you decide to implement the decoder in a programmable logic device, or PLD, with a worst-case propagation delay time of 15 ns over all conditions.

Referring back to Fig. 12.3, then the 50 ns that we had now becomes 35 ns because we've just lost 15 ns decoding the upper address bits and then enabling the proper chip-enable pin on the SRAM. Thus, we either must use faster SRAMs or add an additional wait state.

Here's where it gets tricky. It might work as is for the prototype circuit that the designer has been working on. If your company also does environmental testing, it may or may not continue to work at temperatures closer to 70°C. If you are really lucky, you have a fast batch of 50 ns SRAMs and it even works at temperature. Then, you go to production and the systems start failing in the field because the production lots you are buying are slower than your prototype batch.

At this point, we can get into a very deep discussion about hardware design strategies and using worst-case numbers rather than typical timing values in a design. Is it a design flaw or a bug if some systems fail while others keep working? Would using worst-case numbers compromise the performance of the system or the desired price point? These are not easy questions to answer. However, trying to find the reason for an intermittent failure is tough and possibly very time consuming. Because the memory system is at the very core of any computer system, the debugging process must start with a full timing analysis, followed by measurements with an oscilloscope or logic analyzer capable of time-interval measurements that are better than what you think you need. In this example, a 100 MHz oscilloscope might not be sufficient, but is likely at the edge of being able to show you the relative timing relationships.

A logic analyzer with 2 ns timing resolution would certainly show the timing relationships, but will not show you any analog effects such as slow edges, ringing, or bus contention problems. You need both instruments to do a really thorough job.

I would also consider doing a differential measurement and compare a unit that is intermittently failing with one that is rock solid and compare signals. Record your measurements, capture screen images, and compare waveforms. If you can, do another differential experiment. Replace the SRAMs on your prototype board with the SRAMs from the production batch and rerun your tests. Compare the waveforms with the previous ones. It is always possible to have a bad batch or a good batch that is markedly different than an earlier one.

The last test I would perform is an extensive memory test at room temperature and over a wider temperature range. The memory test should cover the basics test patterns, 00, FFH, AAH, 55H, walking ones, and walking zeros.[b] Any memory failure should stop the test with some indication of where it failed and what pattern was written and what pattern was returned.

Because this is a differential measurement, you want to change just one, and only one, variable at a time. If you are testing the failing board inside a device, and you are testing your reference board using your bench supply, then the first differential measurement would be to swap the device's power supply with your bench supply. Because most products don't ship with $500 lab power supplies, perhaps that's worth investigating.

If you don't have an environmental chamber at your disposal, then find a multimeter with a thermocouple input and tape the tip of the thermocouple down to one of the SRAMs. Try to close up the chassis to make the test as real as possible. Is the SRAM getting hotter than you would expect? Is the ambient temperature inside the chassis hotter than you anticipated? Repeat the test with your reference board in the chassis and compare results.

All the while, use the best practices that were discussed in previous chapters:

- Write down what test you are going to do and what you hope to find out.
- Record your results.
- Analyze your results and use the analysis to guide the choice of the next test to perform.

As I write this chapter, I've been struggling with a problem with a design I did for a student lab experiment to teach how to use logic analyzers. I would have never even thought to mention this as a potential problem before now. The board uses a PLD to do address decoding and generate wait states for the students to observe how they affect performance.

I bought the PLDs from a reputable vendor who has been a reliable supplier for many years. The first batch of 20 parts were from a different manufacturer, but my programmer could program that part from that manufacturer in the package style I was using. I thought, "No problem," until I tried to program them. Not one would program correctly. I tried a second programmer that the students use. Same result.

[b]Jack Ganssle might have a disagreement with me over the test patterns to use. See Ref. [1].

I ordered a second batch and this time I specified the manufacturer. I received 10 parts. Nine programmed properly and one failed to program. I contacted my supplier and they sent me a replacement. The replacement also failed to program, but this time the error was a mismatch in the electronic ID that the programmer reads from the PLD to make sure the correct part is being programmed. We ordered another 20 parts, and I'll see what happens. My suspicion is that my supplier received a batch of defective or bogus parts.

Why do I mention this? Counterfeiting of integrating circuits has become a very profitable criminal activity in recent years because it is so easy to relabel a part and charge a premium price for it. Perhaps you purchase an SRAM with a 15 ns access time and what you receive are parts with 25 ns access times. If your company purchases parts through distribution rather than directly from the manufacturer, now you have to debug the supply chain.

My point here is that debugging a hardware issue may encompass a much wider set of circumstances than you initially thought. Ideally, you'll zero in on a set of test conditions that causes the memory system to consistently fail. Then you are almost home. There is a light at the end of the tunnel.

Dynamic RAM

Unlike SRAM, dynamic RAM, or DRAM, gets its name from the fact that it cannot be left alone and be expected to retain the data written into it. When we examined the SRAM cell, we saw how the positive feedback between the gates making up one bit of storage forced the data bit into a stable state unless some external action (data being written) was taken to flip it.

A DRAM cell has no feedback to maintain the data. Data is stored as charge trapped in a capacitor that is part of the bit cell. At temperatures well above absolute zero, such as the commercial temperature range of 0–70°C, the stored charge has thermal energy and can leak away from the capacitor over time. The mechanism to restore the charge is through a *refresh cycle*. Refresh cycles are a special form of access that appears to be a data read, but the only purpose is to force the DRAM circuitry to restore the charge on a bank of capacitors at the same time. Typically, every bank in the DRAM array must be refreshed ever few milliseconds.

The refresh operation is accomplished by interleaving the refresh cycle with regular memory accesses. Some refresh cycles can be synchronized with the processor clock and are built into the regular bus cycle. The trusty old Z80 CPU had a refresh cycle woven into its regular bus cycles, so memory access never suffered if memory needed refreshing when the processor was trying to access it. When processors don't have built-in refresh capabilities, external hardware such as refresh controllers must be included in your design to handle refreshing and bus contention between the refresh housekeeping requirements and memory access. On PCs, this function was taken over by the Northbridge. This chip connects the CPU to the memory system and provides the proper timing interface to manage the system with a minimum of bottlenecks.

Another major departure in the design of DRAM versus SRAM is how memory is addressed. This comes about because of the relative capacity of DRAM chips and SRAM chips. Today, the leading edge in DRAM technology is 32 Gbits. For SRAM technology, the leading edge is 32 Mbits. That's a factor of 1000 in memory density. This is important because, depending on the organization of the chip, whether it is by 8, by 16, or by 32, all those memory locations require address lines between the processor and memory.

In order to reduce the number of address lines, DRAM chips have evolved a number of strategies. The first technique that DRAMs developed was to split the address into two parts, a row address and a column address. These names are associated with the internal memory array of DRAM cells. If you picture a two-dimensional matrix of DRAM cells, each row of the matrix can be uniquely identified by an address and each column of the matrix can be uniquely identified. The intersection of the row number and the column number uniquely identifies each cell in the matrix of data bits. To create a 16 Gbit DRAM that was organized as 2G by 8 bits, you would need to have eight matrices, each one holding 2 Gbits of data. Each of the 8 matrices would be organized as a 2^{16} by 2^{15} array, or 64 K by 32 K. Depending on how the rows and columns are organized, we could have a 16-bit row address and a 15-bit column address.

A simple DRAM memory access requires that the row address be presented to the DRAM, and *Row Address Strobe* ($\overline{RAS}$) input is asserted to lock in the row address. Next, the column address is presented to the DRAM and the *column address strobe* ($\overline{CAS}$) is asserted. This completes the addressing operation and data can now be written or read.

Fig. 12.4 is a timing diagram for a simple DRAM read cycle [4].

All the I/O signals are the same as for the SRAM, with the exception of the RAS and CAS strobe inputs to the DRAM. It should be obvious that there are complex timing relationships between these signals and designing a DRAM controller is not for the faint of heart. Fortunately, microcontrollers and external

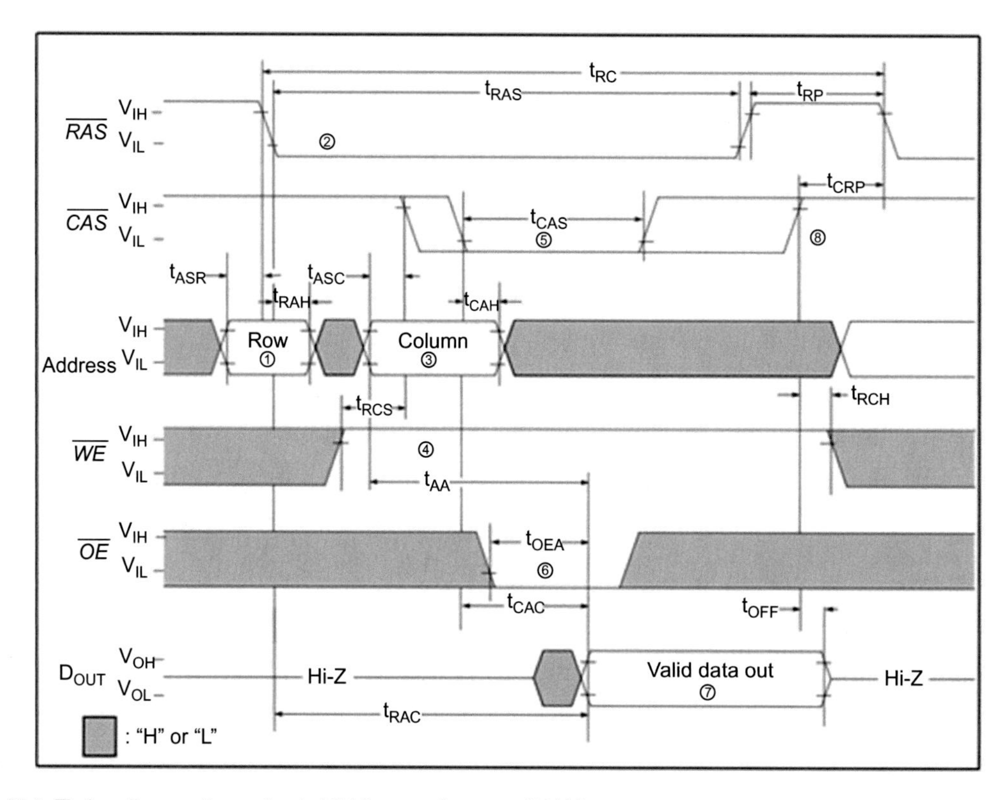

Fig. 12.4 Timing diagram for a simple READ operation on a DRAM.

support chips have greatly simplified the interfacing to DRAM, but the possibility always exists for erroneously programming the internal registers of the DRAM controllers. Here's where a fast oscilloscope can be a very valuable tool to use to examine signal timing and signal integrity of DRAMs.

It may seem that DRAM should be slower than SRAM because of the added overhead of refreshing and the need for RAS and CAS. Actually, for many operations, the DRAM can be faster than the SRAM. While the first DRAM access might be slower, subsequent accesses are incredibly fast because modern DRAM technology has evolved hand in glove with modern CPU architectures, in particular, on-chip program and data caches.

The normal DRAM operating mode is to set up the first address of an access and then issue successive clock pulses that bring in subsequent sequential data on every phase of the clock. We call this *double data rate* or DDR DRAM. This DRAM is synchronous with a clock, which may be the internal clock of the processor or a derivative of that clock. This "burst" access is designed to be very compatible with the design of the on-chip caches.

The operation goes something like this. The processor attempts to access a memory location and the cache-controller looks for that address within the address or data cache. If it determines that it is not in the cache, then the address goes out to external memory. After a number of clock cycles to set up the memory transfer to the cache, the data moves from the DRAM memory on alternate phases of the clock until one row of the cache, perhaps 32 or 64 bytes, is filled. Notice how the size of the burst matches the size of the cache storage region. The burst transfer will occur even if the processor only needs a single byte of data from memory.

While this creates a really tight bonding between the memory and the processor, there is a dark side to this architecture. Suddenly, a new dimension has been added to the ways in which a program could, for no apparent reason, start to run very slowly, perhaps by a factor of 10 or greater, yet the hardware is functioning perfectly well.

Depending upon the nature of the algorithm you are running, the performance of the cache may vary dramatically. This is called the "cache hit ratio," which is defined as the ratio of the number of accesses to a cache and the number of accesses where the cache held the instruction or data the program was looking for. So, if the required data was already in the cache 9 times out of 10, then the cache hit ratio is 90%.

But suppose we craft an algorithm in such a way that the processor is executing repetitive code in a way that each time, part of the data loop is never in the cache and must be fetched from memory. We already saw that the burst is fast but setting up the burst is slow. If the processor is continually requiring cache fills, then the overall performance will fall off a cliff. This is generally called *thrashing* and refers to the constant need to fill and then refill the cache.

To add to the complexity, we also have to account for the effect of the CPU execution pipeline. Modern processors have deep pipelines because instructions can be complex, and the time required to decode and execute the instruction is greater than the 250 ps long clock period of a 4 GHz system clock. Some modern pipelines are more than 20 stages long. While this is not a problem as long as the instructions are all in a long line, a loop will cause the pipeline to be flushed and refilled. Now we have three interacting systems, the DRAM, the cache, and the pipeline.

Unfortunately, I can't give you a simple process to debug poor performance. However, a lot of engineer-centuries have gone into optimizing this system at the PC level. It involves close cooperation between the processor companies, the DRAM manufacturers, the support chip manufacturers, and the compiler designers with a common goal of optimizing the overall performance of the system.

The takeaway here is that if you are using DRAM in an embedded system, unless you are just using a PC as an embedded controller, then these are issues that you may well have to deal with. We discussed the EEMBC benchmark consortium in Chapter 6 on processor performance issues. These realistic industry-specific benchmarks provide a baseline for looking for performance issues with your system.

Soft errors

As engineers, we are comfortable fixing problems that are reasonably deterministic, even if the circumstances that cause the problem are extremely rare and very infrequent. But what if the problem occurs for reasons that we can only guess at and generally attribute to, "it came from outer space?" Imagine telling that to your manager. That's right up there with, "The dog ate my homework."

But it is true. The first source of soft errors was traced to cosmic rays from space. These energetic particles are constantly bombarding us many thousands of times per second. Most just pass right through us and most of the planet, and keep on going. Every once in a while, a cosmic ray will interact with matter. In the case of a DRAM, this is the capacitor that is holding the charge for the bit cell. If the energetic particle hits just right, it can produce a shower of electrons that can change the value of the bit stored in the cell. That is a soft error.

Texas Instruments [5] also cites alpha particles emitted from impurities in the silicon as a source of soft errors, along with the cosmic background neutron flux that is present at low radiation rates at sea level and much higher rates at the flight altitude of aircraft. They also note that alpha radiation can be minimized by the use of ultralow alpha (ULA) materials, but because it is very difficult to shield materials from neutrons, a certain level of soft errors is inevitable.

A white paper by Tezzaron Semiconductor Company [6] concluded that in a 1 Gbyte DRAM memory system, there is a high probability that you can expect a soft error every few weeks, and for a 1 Tbyte memory system, that reduces to one soft error every few minutes. Fortunately, we typically find 1 Tbyte memories in big server systems that would use memory that includes extra bits per byte for on-the-fly error correction. Quoting Tezzaron's conclusion,

> *Soft errors are a matter of increasing concern as memories get larger and memory technologies get smaller. Even using a relatively conservative error rate (500 FIT/Mbit), a system with 1 GByte of RAM can expect an error every two weeks; a hypothetical Terabyte system would experience a soft error every few minutes. Existing ECC technologies can greatly reduce the error rate, but they may have unacceptable tradeoffs in power, speed, price, or size.*

Way back in the early days of the PC when DRAM "sticks" first hit the mainstream, most DRAM modules contained nine DRAM chips rather than eight. The extra bit was a parity bit. Assuming that you've set up your BIOS to check for ODD parity, the hardware will count the number of 1 bits in each byte written to memory and if the number of 1 bits is an even number, the parity bit is set to 1. If the number of bits in the byte is an odd number, then the parity bit is set to 0. When the system reads a memory location, it computes the parity of each byte read back on the fly. If the parity computes to EVEN, then a memory error occurred.

I can remember my PC suddenly freezing with a "memory error" message, but I couldn't tell if it was due to a cosmic ray, a random error, or my PC acting up. So, I rebooted and moved on. If I lost the work I was doing, I muttered some expletives and resolved to save my work more often.

Debugging these errors is nearly impossible unless it happens often enough to devise a debugging test. Perhaps running extensive memory tests for hours or days and not finding an error would at least give you confidence that the hardware is stable and reliable.

The Tezzaron analysis goes on to say that DRAM today is less susceptible to cosmic ray-induced soft errors because while the transistor bit cell has been shrinking, the size of the stored charge needed has not been dropping at the same rate, so the noise margin for error has improved. However, cosmic rays can still play havoc with dense memory systems.

Recent work [7] compared the soft error rate in 40 nm commercial SRAMs by measuring the error rate on a mountaintop and at sea level over the course of several years. The results correlated with the measured neutron flux at both locations. While the root cause mechanisms for SRAMs and DRAMs may be different, studies assert that with today's finer geometries in both SRAMs and DRAMs, soft error rates are roughly comparable. RAM-based FPGAs may also be susceptible to bit flipping and, as one could imagine, the results can be disastrous if the hardware suddenly goes belly-up.

As I discussed in a previous chapter, but is relevant here as well, was the experience of a former colleague who went to work for a supercomputer company. He told me about this experience they had with a new system under development that made heavy use of FPGAs. The FPGAs were mounted on a hermetically sealed printed circuit board and each FPGA sat under a nozzle where a high-pressure stream of Freon refrigerant was pumped onto it to cool it.

The engineers noticed that the electrical noise levels generated in these parts due to the way that the parts were being driven were sufficiently high to flip bits in the configuration memory of the FPGA, even though the parts were being used within their vendor's parametric limits. It wasn't until the FPGA vendor's engineers came out and saw it for themselves did the company realize that they had a design issue with their parts.

SRAM manufacturers are aware of the issue and are directly addressing the problem in their current product offerings. Cypress Semiconductor Corp. expressly discusses the addition of on-chip error code correction (ECC) below 0.1 FIT/Mbit [8].[c] With the addition of on-chip soft error correction, Cypress claims a soft error rate below 0.1 FIT.

In addition to on-chip error correction, semiconductor manufacturers are utilizing a new CMOS geometry called FinFET [9]. FinFET is a three-dimensional structure, as opposed to the traditional planar geometry of the MOS transistor. With FinFET, the gate and the gate oxide are wrapped around the fin-shaped source to drain channel.

Villanueva [10] calculated the soft-error rates for a 6-transistor (6T) FinFET SRAM cell using 20 nm geometries versus a 6T Bulk Planar SRAM cell using 22 nm geometry and found the FinFET to have a FIT value that was two orders of magnitude lower than the bulk planar geometry.

I hope that this discussion didn't cause your eyes to glaze over. If it did, I apologize. I learned a lot by researching it, and I wanted to share my findings. I think the takeaway here is that soft errors are a fact of life in both DRAM, SRAM, and potentially FPGAs as well. Because they are relatively impossible to trace in any traditional debugging protocol one could imagine, the best debugging scheme is to minimize the probability that a soft error will cause catastrophic damage. For systems that are not real time or mission

[c] FIT is an acronym for failures-in-time and 1 FIT = 1 failure per billion hours per chip (https://en.wikipedia.org/wiki/Soft_error).

critical, this may be a minor nuisance and might result in a reboot caused by a watchdog timer resetting the device. If the bit is in a data region of a printer, you might see an errant character being printed. You grumble a bit and reprint the page.

Because SER-manifested behavior[d] can be masked as a software bug, it may be impossible to distinguish what caused the glitch. Of course, in the end, you should be able to determine if the root cause was a software bug, so you might conclude that it was caused by a soft error because that's the only remaining possibility, no matter how low the probability of an occurrence.

For a mission-critical system where human life is at stake and there is a higher than normal probability of a soft error, such as in an aircraft's avionics system, then you know all about this phenomenon and I'm not telling you anything you don't already know. You already build redundancy into your system to prevent these kinds of events.

Mainstream vendors will share their SER measurements with their stakeholders, and you can decide which parts to purchase in order to minimize the potential risk of a SER failure.

Jitter

Jitter is another form of soft error. It is the inevitable uncertainty in the way that electronic devices switch. We often discuss jitter under the general heading of signal integrity and there are a myriad of papers, textbooks, and courses built around measuring and predicting signal integrity in all types of communications networks.

For our purposes, because this chapter is looking at soft error in RAM systems, it is important when trying to find these elusive bugs that we are able to validate our noise and jitter margins. Keysight [11] discusses techniques for validating the integrity of DDR 4 memory systems using oscilloscope and logic analyzer-based measurements. Quoting the Keysight article,

> *Signal integrity is crucial for reliable operation of memory systems. The higher clock rates of DDR4 memory cause issues such as reflections and crosstalk, which cause signal degradation and logic issues.*
>
> *With an oscilloscope, displaying a captured waveform as a real-time eye (RTE) offers insights into jitter within serial data signals. By showing when bits are valid (high or low), the RTE provides a composite picture of physical-layer characteristics such as peak-to-peak edge jitter and noise.*

[d] SER = Soft Error Rate

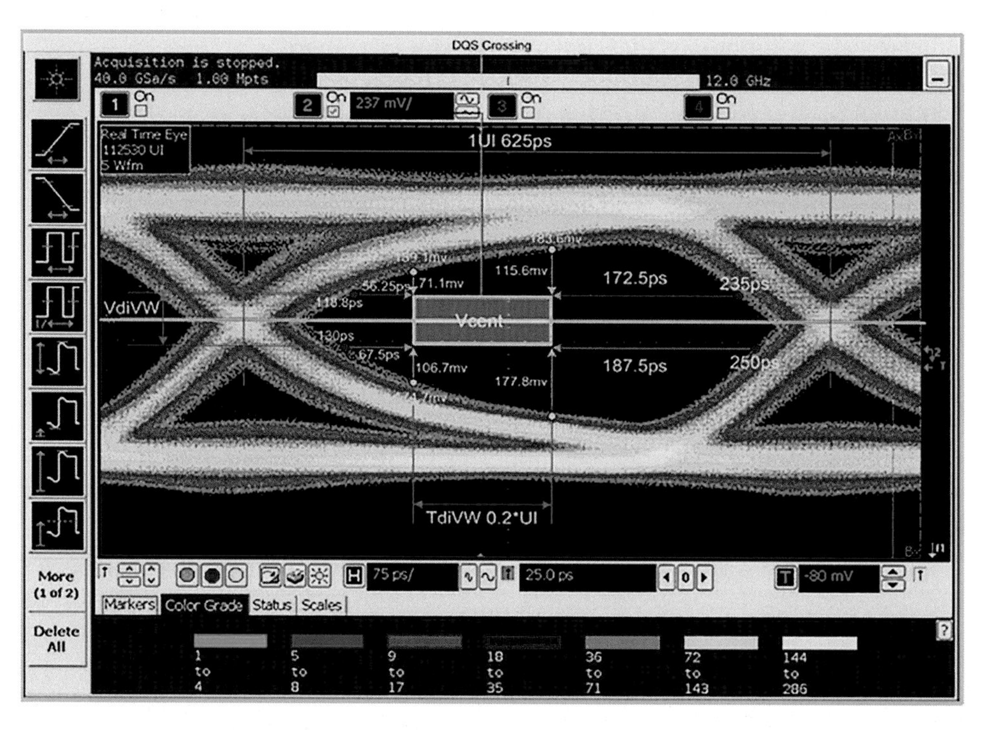

Fig. 12.5 Eye diagram mask testing ensures that signals are not violating the bounds of the mask, where jitter and errors can occur. Courtesy of Keysight Technologies.

> *Getting full insights into the data-valid window and expectations of bit failure requires measurements of the worst-case margins in timing (tDIVW) and voltage (vDIVW). This is done using eye diagram mask testing.*

The eye diagram referred to in the above quote is shown in Fig. 12.5.

Software-based memory errors

We've discussed software-based errors in memory in earlier chapters and there is no need to have a second in-depth discussion about them. However, it would be valuable to summarize some of the most common software bugs that can corrupt memory and crash a system or result in bad data. Fortunately, memory errors due to a flaw in the algorithm should be deterministic, if rare, and there are methods (previously discussed) to ferret out these bugs.

Many software errors are difficult to debug, particularly in programs written in C, because C gives the programmer direct access to memory where their code can make all sorts of mischief. This is great for performance and it puts the C code on par with assembly language in terms of efficiency. Typical memory errors include [12]:

- Out-of-bounds array indexes.
- Buffer overruns.
- Dangling heap pointers (accessing a region of heap-allocated memory after the memory has been freed).
- Dangling stack pointers (accessing a pointer to a local variable of a function after the function has returned).
- Dangling heap pointers (dereferencing a pointer when the pointer points to a chunk of heap-allocated memory that has been previously deallocated via the free() function).
- Uninitialized heap memory (heap memory is allocated using malloc() and some or all of this memory is not initialized before it is read).
- Uninitialized local variable (a local variable of a function is not initialized before it is read).
- The use of pointers cast from incorrect numeric values.
- Vulnerabilities in coding that would allow data transfer overflows to overwrite other variables.

Fortunately, there is a rich set of software tools available from vendors that can be used to uncover these software defects if you take the time to learn them and use them before you deploy your software in your target system. And, as we've previously discussed in how to use the tools of real-time debugging, there are ways to find these flaws in software running in the target system.

When we add an operating system, as we've discussed, other memory defects may be introduced due to the interactions between the RTOS kernel, the hardware devices, and the separate tasks being simultaneously executed.

Concluding remarks

I'm writing this book on a desktop PC I built during the summer of 2019 for no particular reason other than I'm a computer geek at heart and I love to have the latest hardware. The system has an AMD Ryzen processor, 32 Gb of DDR 4 memory, and a 1 Tb solid-state drive. I mention this because I remember the first computer I ever built, a 6502-based system with 64 Kb of RAM using 1 K $\times$ 1 bit SRAM parts that I wire-wrapped together. I stored my programs on audio cassette tape. But it worked. I played games and wrote programs in Basic and assembly language and that computer started me on my path to this one.

In the years between then and now, I built or purchased more than 20 computers and each one was better than the previous one. My computer today has 500,000 times more RAM memory than the first 64 Kb system. In fact, I honestly considered buying four 16 Gb DRAM memory sticks just so I could have 64 Gb of RAM, or 1 million times more RAM than my first computer.

As we've discussed in this chapter, RAM is getting faster and smaller. Semiconductor technologies are pushing the limits that earlier device physicists thought we could never exceed. The FinFET technology is a good example of that. The dark side of the Force[e] is that with the shrinking geometries comes more soft errors. These errors just occur and may occur so infrequently as to be impossible to analyze and correct. The only way to "debug" them is to accept that they are bound to occur with some probability and prepare a strategy against them.

For software, it would involve more defensive programming with rigorous bounds checking to raise the confidence level that the data values being recorded, transmitted, or used in calculations are sane ones.

For hardware, we already know to utilize watchdog timers to guard against an errant programming bug. I hope some of the debugging techniques described in this chapter will provide some insight for how to approach resolving memory issues.

Also, because this is the last chapter, a few more general remarks might be in order. When I was thinking about writing this book, I had two goals in mind:

1. Teach students and new electrical engineers the process of how to find and fix defects in a timely manner.
2. To make the neophytes as well as the seasoned pros, aware of the vast array of literature written by equally seasoned pros, on how to take maximum advantage of the tools that their companies offer.

I've written many of these articles myself in my employment history with HP, AMD, and Applied Micro Systems. My whole professional career has been deeply embedded in test and measurement, with a little materials science at the front end and education at the back end.

Manufacturer's applications notes and white papers are treasures of great information about getting the most out of their tools and on ways to solve gnarly problems. Yes, there is marking fluff

[e]I couldn't resist the reference to Star Wars because the latest release is the final one in the series.

embedded in the app note, but there is great information there as well. With the Internet at your fingertips, any app note or white paper is a few Google searches away.

Thanks for reading this.

References

[1] J. Ganssle, Testing RAM in Embedded Systems, The Ganssle Group, http://www.ganssle.com/testingram.htm, 2009.

[2] D. Gerke, B. Kimmel, EDN Designer's Guide to Electromagnetic Compatibility, second ed., Kimmel Gerke Associates, Ltd, September, 2002, p. 54. ISBN 10: 075067654X.

[3] R.S. Khandpur, Printed Circuit Boards: Design, Fabrication, Assembly and Testing, McGraw-Hill, New York, 2006, p. 164.

[4] http://www.ece.cmu.edu/~ece548/localcpy/dramop.pdf.

[5] Soft error rate FAQs, Texas Instruments, http://www.ti.com/support-quality/faqs/soft-error-rate-faqs.html.

[6] White Paper, Soft Errors in Electronic Memory, Tezzaron Semiconductor, https://tezzaron.com/media/soft_errors_1_1_secure.pdf, 2004.

[7] J.L. Autran, D. Munteanu, S. Moindjie, T.S. Saoud, G. Gasiot, P. Roche, Real-time soft error rate measurements on bulk 40 nm SRAM memories: a five-year dual-site experiment, Semicond. Sci. Technol. 31 (11) (2016) (Special Issue on Radiation Effects in Semiconductor Devices, IOP Publishing Ltd), https://iopscience.iop.org/article/10.1088/0268-1242/31/11/114003.

[8] https://www.cypress.com/products/asynchronous-sram.

[9] Samsung Electronics Co., LTD, Radical Innovation to Push the Limit for Greater Speed and Efficiency, https://www.samsung.com/semiconductor/global.semi.static/minisite/exynos/file/technology/FinFETProcess.pdf, 2018.

[10] M. Villanueva, Analysis of Soft Error Rates for Future Technologies, Universitat Politècnica de Catalunya (UPC), Facultat d'Informàtica de Barcelona (FIB), https://pdfs.semanticscholar.org/c052/8c02f566d211f9bd90b7c1d3703256fad053.pdf, March, 2015.

[11] Application Brief, Physical Layer Validation and Functional Test of DDR4 and LPDDR4 Memory, Keysight Technologies, June, 2015. 5992-0783ENDI.

[12] White Paper, Finding Bugs in C Programs with Reactis for C™, https://reactive-systems.com/papers/r4c_test_tool.pdf, 2002-2011.

Index

Note: Page numbers followed by *f* indicate figures, and *np* indicate footnotes.

F

G

H

I